CHINA'S RUSSIAN PRINCESS:

THE SILENT WIFE OF CHIANG CHING-KUO

by Mark O'Neill

Contents

CHAPTER 1

From the Soviet Union with Love

The life of Faina Ipat'evna Vakhreva (蔣方良) changed forever one freezing night in the winter of 1933 on a street in Sverdlovsk (now Yekaterinburg), on the natural boundary between Europe and Asia. She was walking home with a burly Russian man; his attentions were becoming increasingly unwanted. A young Chinese man was walking home from the late shift at the factory where both she and he worked. He saw what was going on, walked up to the man and demanded that he leave the young lady alone. Seeing a small, puny Asian in front of him, the Russian took no notice. Without a thought, the Chinese threw a punch that sent the tall Russian to the ground – and the young lady was able to get away.

That was the start of a romance between Faina and the young Chinese, Chiang Ching-kuo (蔣經國 , CCK), that was to last 55 years. It would take her away forever from her motherland and sister Anna, who brought her up. Faina never returned home, not even once, after moving to China four years later. Aged 23, Chiang was deputy supervisor of a machine workshop at Ural Heavy Machinery Factory (Uralmash) in Sverdlovsk. Faina, 17, had just graduated from its technical college and

Faina in traditional costume (Courtesy of "Academia Historica")

been sent to work as a turner at the plant. While they first met in the factory, it was this late-night bravery that endeared the young Chinese to Faina and persuaded her to abandon her Russian boyfriend.

But how did these two young people come to meet in this city thousands of kilometres from both their homes?

City of Catherine I

The city of Yekaterinburg was called Sverdlovsk between 1924 and 1991. It is on the Iset river east of the Ural Mountains, in the middle of the Eurasian continent. Tsar Peter the Great named it after his wife Catherine, who took over the throne after his death in 1725. She built the Siberian route, the main road of the Russian Empire, through Yekaterinburg. It became the most important city connecting the rich resources of the region to the rest of the country. The trans-Siberian railway was later built through the city. On November 24, 1723, six days after the official foundation of the city, a giant iron-making plant opened there; it was the biggest in the world at the time. Yekaterinburg became one of Russia's first industrial cities.

Following the Bolshevik Revolution of October 1917, the new rulers took Tsar Nicholas II, his wife, and five children to Yekaterinburg and imprisoned them in the city's Ipatiev House. Fearful that the approaching anti-Communist forces would rescue them, their captors executed the whole family in the early hours of July 17, 1918. Half a century later, in November 1981, the Russian Orthodox Church Abroad canonised the Tsar and his family as "new martyrs"; in August 2000, the Moscow patriarch canonised them as "passion bearers".

On the site of the killing, today stands the Russian Orthodox Cathedral on the Blood. Built to commemorate the Romanov family, it was consecrated on June 16, 2003.

In 1924, the city was renamed Sverdlovsk, after Bolshevik leader, Yakov Sverdlov (1885-1919), a close ally of Vladimir Lenin. The new government nationalised the city's industrial firms and set up its first university, as well as a polytechnic and a medical school. It developed the city as a centre of heavy industry and military production; it created giant new factories, especially in machine-building and metal-working. One of them was Uralmash, short for Ural Heavy Machinery Plant. It manufactured blast furnace equipment, cranes, presses, rolling mills and sintering machines for the mining and metallurgical industries of the Urals and Siberia; it began operations in 1933.

It is in this plant where CCK and Faina were working. People flooded in to the city to work in these new factories; by 1939, its population reached 426,000, more than triple the 136,000 of 1926. By the end of the 1930s it had 140 industrial enterprises, 25 research institutes, and 12 higher education institutes. During the Soviet period, Sverdlovsk was closed to foreigners. The government considered its industrial and military production a matter of national security. Soviet leader Mikhail Gorbachev only lifted this ban in 1991. After the dissolution of the Soviet Union that year, the city regained its historic name of Yekaterinburg on September 4, 1991.

Exile and poverty

Faina was born on May 15, 1916 in Orsha, Belarus and lived there for the first years of her life. Her father was a railroad track lineman. Orsha is in the Vitebsk region, on the fork of the Dnieper and Arshytsa rivers, in the eastern part of what is now Belarus. It is one of the country's oldest towns, with a history of over 1,100 years. In the 16th and 17th centuries, it was the scene of numerous battles and repeatedly destroyed; from the 16th to the 18th centuries, it was a famous religious centre, with dozens of Orthodox, Protestant and Catholic churches and religious orders. It also had a large Jewish population.

In 1772, Orsha became part of the Russian empire. During his invasion of Russia, Napoleon Bonaparte's army captured the town in July 1812; in November that year, the retreating French soldiers burnt it down. In the second half of the 19th century, it developed into an important transport hub, with roads, railways and river boats. By 1900, its firms made flax, leather, bricks, beer; there were mechanical and iron foundries. In 1904, it had 14,784 people, of whom about half were Jewish. In the first half of the 20th century, Jewish people were the third largest ethnic group in Belarus.

During the First World War of 1914-1918, Belarus suffered greatly. The Tsarist government deported people of German and Czech origin to the interior of the empire for fear that they were spies; the deportations worsened during a major retreat of the Tsarist army in 1915. The Jewish population was also moved eastward and the military conducted anti-Jewish pogroms. In total, more than 1.4 million people were deported from Belarus. During battles there in 1916 and 1917, both sides used heavy bombardment and poison gas. Under the Treaty of Brest-Litovsk signed in March 1918, the new Bolshevik government ceded control of nearly all of Belarus to Germany. The German army occupied Orsha between February and October 1918. Then the area became the scene of battles between the Polish state reborn after World War One and the new Soviet government (the Russo-Polish War of 1919-20). Faina's family was Russian Orthodox.

Faina's family lived through these tragic times. In about 1921, they left Orsha and moved to the relative safety of Yekaterinburg, 2,300 kilometres to the east.

During her childhood, Faina lost both her parents. She was brought up by an elder sister Anna, who became a substitute mother; Faina lived in dormitories and worked as a milling machinist, producing small parts. Life was difficult; with no help from relatives, the two sisters had to rely on each other. During the Russian Civil War from 1917-1922, living conditions deteriorated; the streets were lined with beggars and the starving. The Communist Party won a savage war against White Russian armies supported by 11 foreign nations; between eight and 12 million were killed.

In 1929, at the age of 16, Faina went to a technical school under Uralmash where she learnt to cut metal. She studied there for three years. Since the giant plant was being built during this period, she and the other students would go to work on the site in the evenings after classes. Living conditions in the Soviet Union were harsh; most goods were in short supply and many rationed. For the employees, belonging to a major national project like Uralmash had many advantages – its canteens were well supplied and they were paid regularly. It provided apartments and dormitories for them to live in, as well as social and sports facilities. The government wanted the plant to run efficiently and its staff to be adequately fed and physically fit. Faina joined the Communist Youth League.

It became possible to draw a portrait of Faina during this period only after the collapse of the Soviet Union in 1991. For the first time, people from Taiwan were able to go to Yekaterinburg and meet their former classmates and colleagues; many had stayed at Uralmash for their entire working lives. One was Tatiana Alexeevna Kaielina. She said that she and Faina were close and shared a common interest in ice-skating; since Tatiana was accomplished in this sport, she became Faina's teacher. "Faina also liked swimming and going to the cinema with her friends to watch films. On holidays, they went cycling. She enjoyed everything in life," she told Zhou Yu-kou (*Chiang Fang-liang and Chiang Ching-kuo*, page 57).

Another was Maria Semenovna Anikeeva; she retained a photograph of Faina who gave it to her in 1935 and wrote on the back "no matter how far away we are, the feeling between friends lasts for ever". [*Chiang Fang-liang and Chiang Ching-kuo*, Zhou Yu-kou, page 59]

The young woman they describe was warm and enthusiastic. She had overcome the hardship of being an orphan and threw herself into her work and social life; she enjoyed everything life had to offer. Another characteristic was her neat and clean appearance, difficult to maintain

Faina (Chiang Fang-liang, right) and her elder sister Anna who brought her up

in the dust and dirt of a heavy industrial plant. Slender and attractive, she and Tatiana both had many suitors.

Extraordinary odyssey

Chiang Ching-kuo's (CCK's) odyssey to Uralmash was even more remarkable. The world he came from could not have been more different to that of Faina. He was born on April 27, 1910 in Xikou, Zhejiang province (浙江省溪口鎮). His father was Chiang Kai-shek (CKS), son of a landowner and a prosperous businessman who sold salt and wine. CKS was one of the few Chinese to attend Shimbu Gakko (振武學校), a preparatory school for the Imperial Japanese Army (IJA); he then served as a private in the Japanese army. In Tokyo, he joined the Tongmenghui (同盟會), a revolutionary party that was fighting the Qing dynasty. When he learnt of the Wuchang Uprising in October 1911, CKS rushed back from Tokyo and boarded a boat to Shanghai. He led an attack on Hangzhou; it was successful and he went to hold high positions in the revolutionary army.

While CKS was away fighting, CCK grew up with his mother in the peace and quiet of Xikou, in eastern Zhejiang; it was untouched by the chaos that enveloped much of the new republic. Just before he was six, CCK went to study at a traditional school in his hometown. He was a happy, obedient child. In 1920, his father sent him to a larger school in the local county town. Two years later, he moved to a grammar school in Shanghai, staying with his father's second wife Chen Chieh-ju (陳潔如) in her home in the French concession.

Chiang Kai-shek impressed by Soviet Union

In September 1923, Chiang Kai-shek led a four-man delegation to Moscow to buy weapons and see the workings of the new communist state. He was greatly impressed by what he saw, especially the Communist Youth League and the system of political commissars.

China's revolution of 1911 had not created a modern, democratic state – it was a country divided into regions run by warlords and their own private armies; the "central government" was weak and controlled only a small portion of the national territory. It was different in the Soviet Union. During the civil war, the Red Army had succeeded in unifying the world's largest country, 22.4 million square kilometres stretching from the Polish border to the Pacific Ocean. Now the government was attempting to build the world's first Communist state, an experiment never tried before in history; it faced the hostility of the world's richest and most powerful nations.

Like many Chinese, CKS saw much to admire and emulate in this new country. He considered what parts of it he could copy at home. He was also impressed by the fact that, while the western powers clung to their commercial and territorial privileges in China, the Soviet government was willing to help the new republic, especially in the military field.

In January 1924, the Whampoa Military Academy was set up in Guangzhou and Chiang appointed its first director. He invited a Soviet general, General Vasili K. Blyucher, to serve as his chief-of-staff. The young Soviet state began to provide weapons to the Chinese government, including artillery.

In early 1925, CCK graduated from his school in Shanghai and entered the Pudong High School (上海浦東中學). Later that year, his father sent him to a new school in Beijing; there he met members of China's four-year-old Communist Party and visited the Soviet embassy, to watch films and meet diplomats. Like many young Chinese, he was very interested in what he saw and heard.

He asked his father if he could enrol in the new Sun Yat-sen University (中山大學) in Moscow; the Soviet Communist Party had set it up that year to train Chinese revolutionaries. Initially, CKS opposed the idea. But, in October, he changed his mind and approved the proposal; his son was only 15. It was a political decision – CKS wanted the military and diplomatic support of the Soviet Union. He also wanted CCK to mix with the sons and daughters of other KMT leaders who had been sent to Moscow. For a father, however, it was a bizarre choice – to send his son, so young, to a city thousands of kilometres from home where he could not protect or look after him and in a country of which he knew little.

To the centre of the revolution

In late October 1925, with 90 other Chinese students, CCK boarded a Soviet steamer in Shanghai for the journey to Vladivostok. From there, they took the trans-Siberian railway; at Moscow station, they were met by Karl Radek, rector of the new university, who escorted them to the campus. About 80 per cent of the students came from the Chinese Communist Party and only a minority from the Kuomintang. The students enjoyed a standard of life better than many Soviet citizens at that time – for breakfast, they had eggs, bread, butter, milk, sausages, and black tea. The university hired a Chinese cook to make dishes they liked.

Chiang chose the Russian name, Nikolai Vladimirovich Elizarov; this was not by accident. For a time, he lived in the house of Lenin's elder sister, Anna Elizarova-Ulyanova; so he chose Elizarov and Vladimir, the first name of Lenin. "Nikolai" adapted to his new surrounding like a duck to water. He learnt Russian quickly and threw himself into academic and social life; he wrote in the student magazine. In January 1926, he was joined in his 20-student class by Deng Xiao-ping (鄧小平), who was on his way home after work and study in Paris; the two young men walked along the Moscow River and discussed politics. More than 50 years later, they would be the most powerful leaders of the two states of China – but never met again.

In the summer of 1926, CCK had his first girlfriend, Feng Fu-neng (馮弗能), daughter of a Chinese warlord, Feng Yu-xiang (馮玉祥); she was just 15 years old. The two married, a simple process at that time in the new Socialist state; it did not even require registration. He had done well enough in his political studies and learning Russian to be invited to give public speeches. He addressed 3,000 Moscovites on the Northern Expedition which his father had launched with the aim of unifying China. In the summer of 1927, after graduating from Sun Yat-sen University, he applied to return to China – but Stalin refused.

Thesis on guerrilla war

That autumn, CCK was transferred to the prestigious Central Tolmachev Military and Political Institute in Leningrad, the top academy of the Red Army, where he was taught by Marshal Mikhail Tukhachevsky, a famous Red Army general. In January 1928, he became a member of the Soviet Youth League and, later that year, a candidate member of the Soviet Communist Party. In May, his bride, Miss Feng, was allowed to go back to China, together with her brother. The two divorced.

In May 1930, he graduated top of his class, having written a thesis on guerrilla warfare. His personnel file described him as the best student at the academy. He applied to go home – rejected again. He returned to Moscow and was hospitalised because of an illness caused by diabetes; this had been exacerbated by heavy drinking of vodka.

By now, CCK realised that he was not simply a young man studying at a foreign university. He had become a piece in the complex and Machiavellian game of Soviet politics and diplomacy. Stalin had two allies in China – the Kuomintang government and the Communist Party set up in 1921; his support for one or the other changed constantly. In this game, as the son of China's president, CCK was an important piece. He had become a hostage of Stalin; he was more valuable to him as a prisoner in the Soviet Union than if he were allowed to go home. Stalin used him to pressure his father to do what he wanted.

Another obstacle to his returning home was the Chinese Communist Party and its branch in Moscow; it too wanted to use CCK to achieve its objectives in China. The chairman of the CCP branch at Sun Yat-sen University from April 1926 was Wang Ming (王明). Wang had arrived in Moscow on the same train as CCK. He quickly mastered Russian and Marxist-Leninist theory and became close to the vice-president of the university, an influential member of the Soviet Communist Party. In 1926, Wang became the representative of the Chinese party in the Comintern (the Communist International) in Moscow. He intensely disliked CCK for his political views and his influence over the other Chinese students. In deciding whether to allow CCK to go home, Stalin had also to consider the views of Wang Ming and the Chinese Communist Party.

Complicating things further was a deadly split in the Soviet Communist Party between Stalin and Leon Trotsky, the founder and commander of the Red Army that had won the Civil War. The split between the two developed from 1923. At Sun Yat-sen University, most of the teachers favoured Trotsky, as did their Chinese students, including CCK. In 1927, CCK wrote a poster in praise of Trotsky and stuck it on the wall of the university; with a group of classmates, he went to visit him. But, in October 1927, Trotsky was driven out of the central committee of the party; he was expelled from the Soviet Union in February 1929. On August 20, 1940, he was attacked in Mexico

City by a Soviet agent with an ice axe and died the next day; the agent was acting on Stalin's order. In January 1928, CCK had to renounce his support for Trotsky – but it was political ammunition his enemies could use against him.

Meeting Josef Stalin

After his graduation from the Leningrad Military Institute and his recovery from illness, CCK was sent to the Dynamo Electrical Plant in Moscow as an intern. Each day he did eight hours of hard labour as a machine tool operator, for a monthly wage of 45 rubles; in the evenings, he studied engineering at the Lenin International School. Food shortages were so severe that he sometimes went to work without breakfast. Then, suddenly, in the autumn of 1931, he was summoned to a meeting with Josef Stalin. He wanted to talk to the young man about Japan's recent occupation of northeast China, Manchuria; Stalin wanted the KMT and Communists in China to form a united front against Japan and prevent it from attacking the Soviet Far East.

In early 1932, Stalin suddenly decided to send CCK to the Zhukova collective farm in Korovinsky (now Ryazan), 200 kilometres southeast of Moscow. The first night he slept in the warehouse of a church. Most of the workers were illiterate; they doubted that this bookish city boy would be able to work the land. To prove them wrong, CCK threw himself into the job, exhausting himself in the process. He earned the trust of the workers and a good evaluation in his official file, as someone with leadership skills. Within a few months, he became chairman of the collective. Unfortunately for CCK, Wang Ming, the Chinese Communist Party representative in Moscow, was lobbying hard against him. He argued that CCK should be sent further away from the capital, to diminish his influence on the Chinese students.

Siberian exile

So, in October 1932, CCK was sent to Sverdlovsk, 1,800 kilometres from Moscow. After he arrived there, he fell ill again and was hospitalised for 25 days, the third time in three years he had needed such treatment. After his recovery, in January 1933, he was sent to Altai, on the southern border of the Soviet Union, next to Kazakhstan, Mongolia and Xinjiang. He described his nine months there as a "Siberian exile". CCK worked next to professors, students, aristocrats, engineers, rich farmers, and robbers. Each had an "unlooked-for, unexpected misfortune which had sent him into exile," he wrote in an account of his life there, published in China in 1947 (*My Life in the Soviet Union* by Chiang Ching-kuo).

This was probably a "gulag", a forced labour camp. According to Soviet archives that became available after the collapse of the Soviet Union in 1991, about 800,000 people worked in these camps and a further 300,000 in labour colonies in 1935. The inmates were petty criminals, POWs of the Russian civil war, officials accused of corruption, sabotage and embezzlement, political enemies, dissidents and other people deemed dangerous to the state. Many were located in remote regions, like Altai, and given tasks such as building large infrastructure and industrial projects, exploiting natural resources and opening up new land for settlement.

Altai covers an area of 92,600 square kilometres in west Siberia. For more than two millennia from 209 BC, it had been controlled by nomadic, Chinese or Mongol rulers and their states; the most recent Manchu ruler was the Qing dynasty, from 1757 to 1864. From 1864 to 1867, Tsarist armies conquered the region and encouraged Russians to settle there; in the 2010 census, they accounted for 57 per cent of the region's population. For Josef Stalin, it was an ideal place for gulags where he could expel people he did not like, such as political opponents, dissident intellectuals, landlords, rich peasants, members of the nobility and those who had supported the White Russian side in the civil war. It was 3,230 kilometres from Moscow and had no railway. It was a remote and mountainous region mainly dependent on agriculture. For the young CCK, it was hard to imagine a place further from the centres of power and influence where he had spent his first eight years in the Soviet Union. In October 1933, because of his excellent work record, he was allowed to return to Sverdlovsk.

There the Communist Party committee of the Urals district sent him to Uralmash, where he became deputy supervisor of a machine workshop. This was one of the most important factories in the city and a sign he was well regarded by the party. Initially, some of those working under him resented having a foreigner as their superior. But he won them over with his ready smile, easy manner, and good temper. Among them was the young Faina. Their initial impressions were mutually favourable. CCK saw an elegant and beautiful young woman with blonde hair and blue eyes; she saw a warm, lively and extrovert personality. She said later: "He was always smiling". The only Asian in Uralmash and one of the few foreigners in Sverdlovsk, CCK became a minor celebrity in the factory. Like Faina, he had many suitors.

Given the political atmosphere of that time, his success was remarkable. The Soviet Union of Josef Stalin was xenophobic and suspicious of foreigners. It knew that the world's major powers hated the Communist revolution; they feared it would be contagious and spread to their own countries. During the Russian Civil War, 11 nations had sent soldiers to help the White Russian forces fight the

Red Army; among them were Japan, which sent 70,000 troops, and the Republic of China, which sent 2,300. After it occupied Manchuria in 1931, Japan's military considered an invasion of the Soviet Far East, to seize natural resources the country needed for its industrialisation. So an Asian like CCK could easily be suspected of working for Japan. In 1936, the Soviet Union sent home many Chinese migrant workers because it feared their loyalty. In 1937, Chinese who remained in the Russian Far East, as well as ethnic Koreans, were deported to areas of the Soviet Union far from the border for fear that their communities might be infiltrated by Japanese spies.

CCK had succeeded in overcoming these suspicions. After studying and working for eight years in the Soviet Union, CCK spoke Russian fluently. His technical competence, easy smile, good humour and fondness for vodka and dancing made him popular with his fellow workers. Since arriving at Uralmash in 1933, CCK had won the trust and affection of his superiors and colleagues despite his colour and nationality. He started as head of the plant's complaints department and rose to become assistant to the chief of the largest machine shop, Number One; he finally became editor of the factory newspaper *Heavy Machinery*. He mixed easily with the other workers, eating, drinking, telling jokes and dancing with them. He liked Faina more than other lady colleagues but was unsure how to court her, especially since he was her superior. So he engaged the help of a commissar of his workshop, Fyodor Anikeev, to introduce them. Fyodor was chairman of the Communist Youth League in the factory. This allowed him to invite both CCK and Faina to his office, a more private meeting place than the factory floor or canteen.

Rationing and Shortages

Life for CCK, Faina and other residents of Sverdlovsk was harsh, if better than many in the new Soviet republic. The government had designated the city as a major industrial centre. Thousands moved there to work in the new factories and research centres. But the construction of new apartments could not keep pace. Many people lived in frame houses, huts, basements and makeshift dwellings. Only a minority lived in comfortable apartments. Food and other essential goods were in shortage. Those who worked at major enterprises like Uralmash were fortunate; they received cards and coupons enabling them to secure a minimum of food and manufactured goods.

There was also a shortage of doctors and hospitals. Already emerging was the dual-class system that became a characteristic of Communist countries. On top was an upper class nomenklatura of senior party and government officials who received wages several times higher than those of their workers. They had access to spacious apartments, well-stocked stores, hospitals and nursing homes off-limits to ordinary people. Below were workers, farmers and other common people; they soon learnt to disbelieve the claims and promises of official propaganda on radio, television and newspapers.

In response to the lies, shortages and inequality, Russians began to create jokes that became a feature of Soviet rule, in Russia and later in Eastern Europe. Here is one example:

Three men are sitting in a wooden hut in a gulag in Siberia. It is minus 40 degrees Celsius. They have just arrived and are very nervous. But it is so cold that they feel the need for human warmth. One asks another: "why have you been sent here?"

He replies: "I supported Beria (head of the NKVD, the secret police). How about you?"

"I opposed Beria," the second answers. They turn to the third. "And you?"

"I am Beria."

Faina and CCK swimming in a river near Sverdlovsk, mid-1930s

Romance blossoms

Faina and CCK were similar – extrovert, warm and lively. Both enjoyed cycling, swimming, skating and the outdoors; as well as social life, including drinking and smoking. They went on excursions and picnics in the countryside with their colleagues and friends. She was a member of the Communist Youth League and eagerly took part in its activities.

In Sverdlovsk society, both were outsiders – he was a foreigner thousands of kilometres from his family; she was an orphan, with elder sister Anna her only relative. "When they were young, they always laughed a lot," said Chiang Hsiao-yong (蔣孝勇), the youngest of their three sons. "Mother saw Father as a brave hero." (*Biography of Chiang Fang-liang*, Wang Mei-yu, page two). She knew he belonged to an important clan in China – but not that it was the "first family". Nor did she imagine that she would leave her home country and go to live permanently in China. In his memoirs *My Life in the Soviet Union*, CCK described Faina as "his only friend at Uralmash ... She was an orphan. We met after she graduated from a workers' technical school. I was her superior. She was the person who most understood my situation. Whenever I experienced difficulties, she always expressed sympathy and a willingness to help."

In March 1935, Faina and CCK married through a simple registration process; he was 24 and she was 18. Witnesses at the ceremony included her sister Anna and Anna's husband, Fyodor from the Communist Youth League and his wife and another colleague. CCK was earning 700 rubles a month, a large salary by the standards of the time. The new couple took their honeymoon on the Black Sea, a privilege reserved for the best-regarded cadres and workers. They went to Sochi, the city which hosted the Winter Olympics in 2014 and where President Vladimir Putin has a giant palace that cost an estimated US$1 billion.

But, back in Sverdlovsk, the apartment of the newly married couple was small. In 1975, CCK wrote in his diary: "There was only room to fit one bed and one table. We were bothered by smelly insects. We never had a good night's sleep. In a month, it was hard to obtain a bar of soap. In a week, it was hard to obtain a small piece of beef. I and my wife relied on our own strength. We should remember these hardships."

Later in life, he and Faina liked to tell their children stories about their life in Sverdlovsk, to make them realise how privileged was their life in Taiwan. At their new apartment, the couple entertained their friends and fellow workers. CCK enjoyed Russian songs and dancing in the energetic Caucasian style; he was also keen on a Chinese drinking game "hua chuan" (猜拳 , matching numbers), a finger game at which he was very gifted. Guests at their apartment included the director of the Uralmash plant and the regional Communist Party secretary. In December 1935, Faina, then 19, gave birth to their first son Hsiao-wen (孝文); his English name was Alan. CCK was delighted. During the baby's first three months, his parents took turns to feed him at night. Alan was born prematurely; he was physically weak and needed intense care from his parents. This brought them closer together and gave them a special feeling for their eldest son. The family was a haven of warmth and intimacy in a city of bitter winters and political uncertainty.

In his memoirs, CCK gives a flavour of his life in Sverdlovsk in 1935. "After work was finished, I went with four friends to a sports stadium to watch a football game. Recently, I have developed a strong interest in football. After the game, we went to the cultural garden and played games with my fellow workers." On the last day of 1935, the factory organised a New Year dinner attended by more than 1,000 people. "The tables were laden with food and drink. They made everyone think of the hunger of three-four years before. So we were especially happy. On the stroke of midnight, the Communist Party secretary of the region said that 1935 was over and a new year had started. He invited everyone to raise their glasses and drink to a happy and blessed future... Everyone emptied their glasses in one go. On the stage, people sang and danced; it was very noisy. After the revolution, the Soviet Union did not promote the New Year. In recent years, there was nothing to see on January 1, which was a very normal day. This was a sign of the difficulties of life. You could not buy fish, meat or bread; how could you celebrate the New Year? But this year the Soviet government has been urging people to celebrate the New Year festival." He reached home at half past midnight and attended another dinner, for his eight best friends from the factory. "Two weeks ago, our son was born. Because I was very busy at work and did not invite guests. So I decided to host a small dinner on New Year's Eve." He entertained the guests until early morning. After they went home, he could not sleep, so he read letters of good wishes from his former colleagues at the collective farm near Moscow and a fellow student at Sun Yat-sen University. (*My Life in the Soviet Union* by Chiang Ching-kuo, page 63/4)

His promotions in Uralmash were remarkable for a foreigner and an Asian. It would only give the position of editor of the factory newspaper to someone trusted by the Communist Party. The promotions were evidence of how well he had adapted to his new life and won the trust and affection of his superiors and fellow workers. He was allowed to give lectures on international affairs. These promotions, his fluency in Russian, the choice of a Russian wife and his enthusiastic participation in work and party activities all suggest that he did not expect to return to China. By 1936, he was an alternate member of the Communist Party, the final step before full membership.

The memoir of his time in the Soviet Union reads like the work of someone fully committed to its ambitious revolutionary mission; it is a very positive account. That is certainly what Faina thought. She had married a lively and attractive man, with a good salary, who had achieved a great deal during his decade in the Soviet Union. Looking at his skills, connections and charm, she could look forward to a comfortable family life, much better than what she had endured as a child. And this life would be in her own country, with the language, food, friends, and lifestyle with which she was familiar.

The Great Terror

In Stalin's kingdom, CCK was, like everyone else, not free in his words or his actions. He was a hostage both of Stalin and of the Chinese Communist Party's representative in Moscow. He was under the regular surveillance of the NKVD, the secret police. Sergey Ageyev, a historian of the Uralmash museum, said that, in an interview, Faina once said that she and her husband were controlled and followed 24/7 – at work and outside. (Article by James Baron, The Diplomat online news magazine, 28/9/2018)

A biography of CCK by a mainland academic Chen Shouyun (陳守雲，《洞悉蔣經國》, *Know Clearly Chiang Ching-kuo*) quotes from an interview in 1996 by Maria Anikeyeva, a friend and co-worker with Faina in the Uralmash factory. "Faina told me that, in 1935 and 1936, agents of the KGB often came to see her. They warned her 'you must be careful. If you are alone in the apartment with your son, do not allow strange men to enter the house. At any time, Chiang Kai-shek could secretly send people and take you by force to China'."

The political atmosphere worsened in the summer of 1936, with the first public trials of the Great Terror, in Moscow. This was a devastating purge of officials of the Communist Party, the government and the military and wealthy peasants accused of being "enemies of the state" or "enemies of the people". In total, the NKVD arrested 3.8 million people for "counter-revolutionary crimes" and executed over 780,000. They included many of the Party officials whom CCK knew in Sverdlovsk and invited to his apartment, as well as engineers and technical specialists; they were labelled "bourgeois experts". Most terrifying about the Terror was its arbitrary nature and summary justice; those arrested often did not know what they were accused of. The "trials" were short and pre-decided; there was no defence or appeal. A strong admirer of Stalin, Mao Tse-tung used similar tactics to attack his enemies, real or imagined, after he established the PRC in 1949.

As an Asian and one of the few foreigners in Sverdlovsk , CCK was an obvious target for such a political campaign – and so it turned out.

In September 1936, the Ural Communist Party Committee suddenly informed him that he had been dismissed from his posts at Uralmash; it also cancelled his alternate membership of the Communist Party. At a party meeting, someone had accused him of being a "Japanese spy" and a "Trotskyite"; these were most serious accusations. CCK never saw the evidence against him and had no chance to defend himself. There was no court to which he could appeal nor person to whom he could turn for help. With this Kafkaesque decision, his 11 years of hard work and devoted integration into the Soviet system suddenly evaporated into thin air. He could no longer go to the factory, the canteen or the social club. While Faina went to work, he stayed in their small apartment and spent the day reading books and looking after Alan. Fearful of "contagion", his friends did not dare to visit him. This talkative, sociable man suddenly found himself without company, an outcast and untouchable. He could not sleep normally. At any moment, officers of the NKVD could break through the door and take him away. He did not know if each day was his last. In the Great Terror, the charges against him could lead to arrest, imprisonment and even execution. It was an experience he would not forget.

These weeks after the dismissal was his most difficult period in the Soviet Union. The family was suddenly dependent on Faina's meagre salary; fortunately, she rose to the occasion and managed the home capably with their lower income. Hardship and thrift were the life she had always been used to – and the patience to endure it without complaining. She and baby Alan supported him during this terrible ordeal.

Saved

What saved CCK from possible arrest and execution was something that happened 4,400 kilometres away in December 1936 – the Xian incident. On December 12, troops loyal to Marshal Zhang Xue-liang (張學良) arrested CKS near a hot spring resort outside Xian where he was staying. The Marshal demanded, as a condition for his release, that CKS end his military campaign against the Communists and form a united front against the Japanese. For a few days, confusion reigned; anything was possible. Would the government launch a campaign to rescue their president? Would he be executed by his captors? In Moscow, Stalin reacted immediately; he desperately wanted such a united front. His worst-case scenario was a pro-Japanese KMT government in Nanjing; this would allow the Japanese military to deploy its substantial resources in an invasion of the Soviet Far East. Stalin sent a long telegram to the Communist leaders in Yanan saying that CKS was the only person who could lead a united front against Japan. They followed his advice and offered to put their military forces under KMT control; they also promised that, as part of the deal, CCK would be allowed to return home. Under this strong pressure, the

Marshal released CKS and the two sides agreed on a united front.

Back in Sverdlovsk, CCK read about the Xian incident on December 14. He realised at once this was the opportunity he had been waiting for. He wrote a letter to Stalin saying he wished to go home to help in the new united front. A week later, he was summoned to Moscow; senior officials there told him that they looked on the Chinese government with favour and told him to return to Sverdlovsk to await orders. A few weeks later, he was officially informed that he and his family would be allowed to leave. He was very excited and told Faina. Her reaction was more ambiguous; she was happy to end the daily surveillance and threat of being purged. But this was not the outcome she had expected. It would mean leaving behind the sister who had brought her up, as well as all her friends and co-workers in Sverdlovsk. She was going to a country where she knew no-one. She did not speak Chinese – she and her husband talked in Russian. She had no idea of the life that awaited her in this new country, nor its complex culture or traditions. And it was in a war with the most powerful military nation in Asia. Her heart was full of anxiety.

Going home

In February, CCK was summoned to Moscow again, this time with his family and belongings. Their many friends and co-workers came to the station in Sverdlovsk to bid them goodbye. It was bitterly cold. They held a tea party on the platform and drank and danced together. Faina's sister and friends wept and blessed her for her new future. The farewell was a testament to how well the young Chinese had integrated into Soviet society.

The deep emotions of that day were described in a letter which Faina received in 1992 from her friend and co-worker Maria Anikeyeva: "Of our days together, two incidents have left an especially deep impression in my memory. One was the New Year together, in 1935 or 1936. Nikolai (CCK) had brought good liquor from Moscow. The four of us (including Maria's husband) savoured it. That day your son cried very loudly. I remembered holding him and rocked him on the black sofa, to try to please him. The second incident was the day when we saw you off at the city railway station. It was the last time we saw you. There were seven or eight of us. You used a blanket to wrap your son. Everyone was sad. The women wept. The men embraced Nikolai to say goodbye. I thought 'love has such great power. For love, a person can leave their mother country.'" (*Biography of Chiang Fang-liang*, by Wang Mei-yu, pages 121 and 122)

This letter conveys the sense of shock felt by their friends and fellow workers at this sudden departure of Faina, her husband and son. They

CCK, Faina and son Hsiao-wen in China late 1930s

expected the family to be their long-term friends and companions in Sverdlovsk. None knew when they would see the three of them again, especially because they were going so far away. In Stalin's Russia, you could not predict what would happen a week later, let alone the years ahead.

In Moscow, CCK was greeted by his friends in the Red Army. At large meals washed down with vodka, they promised him that their country would do all it could to help China defeat Japan. Stalin also invited him for a meeting to stress the importance of the united front between the KMT and the Communists. In early March, Chinese ambassador Tsiang Ting-fu (蔣廷黻) invited the couple to the embassy for a dinner. In his memoirs, the ambassador described Faina as a beautiful blonde who was extremely shy. No wonder – she had no experience or knowledge of this rarefied political and diplomatic world into which her husband was taking her. The embassy graciously provided CCK with a western suit and his wife with a long, elegant coat and dress.

From Moscow, Father, Mother and son took the trans-Siberian railway for the long journey across Russia. They arrived in Vladivostok, which was cold and forbidding. They boarded a Soviet cargo ship that took them to Shanghai. All they were carrying were the young Alan and their suitcases. None of the three set foot in Russia again during the rest of their lives.

Main sources for this chapter:

The Generalissimo's Son, by Jay Taylor (Harvard University Press, 2000).

My Life in the Soviet Union, by Chiang Ching-kuo (Qianfeng Press, 1947).

2007 article in *Konsomolskaya Pravda* in Belarus on CCK and his wife.

Biography of Chiang Fang-liang, by Wang Mei-yu (China Times Publishing Company of Taiwan, 1997).

Know Clearly Chiang Ching-kuo, by Chen Shouyun (Independent Authors of Taipei, 2016).

International Encyclopedia of WW1 – Belarus, essay by Andrei Zamoiski.

Official website of Orsha city

The Cog that Slipped: Chiang Ching-kuo's Russian Odyssey, by James Baron (The Diplomat website 28/9/2018).

Ekaterinburg, Historical Essays (1723-1998), Report by Tass News Service published in Yekaterinburg 1998.

Chiang Fang-liang and Chiang Ching-kuo, by Zhou Yu-kou (Rye Field Publishing Company, 1993).

Pravda, 9/4/2013, report on Yekaterinburg's bid to host the World Expo 2020.

CHAPTER 2

Joining the First Family, Fighting Japan

The Soviet freighter sailed from Vladivostok to Shanghai. On April 19, 1937, it moved slowly along the Huangpu River. Faina stood on the deck, next to her husband, holding the two-year-old Alan. The two sides of the river were lined with factories and boat yards. Several warships of the Imperial Japanese Navy were at anchor, off the Bund. This was China's industrial and commercial capital. Faina had never seen anything like this before.

At the pier, the three were met by a member of CKS's personal office, the Mayor of Hangzhou and a group of bodyguards. The party went to Shanghai railway station and boarded a train for Hangzhou. CKS did not invite them to the capital Nanjing. The family stayed for several days in Hangzhou before CKS came to see them; the meeting took place in the presidential residence in the city. CCK knelt on the floor and touched his forehead three times in front of his father. After the meeting of Father and Son, CCK introduced his wife and Alan, who were sitting downstairs. Faina met CKS and Soong Mei-ling (宋美齡), CKS's wife, for the first time.

These early weeks were a great challenge for the 21-year-old. She had discovered that not only was she moving to a new and unknown country but had joined its most powerful family. She could not speak Chinese. The contrast between Faina and Madame Soong could not have been sharper. Madame Soong was the most sophisticated Chinese woman in the world; educated at the elite Wellesley College in Boston, Massachusetts, she spoke fluent English and mixed effortlessly with the elites of the West as well as of China. She was rich, self-confident and elegantly dressed.

Faina was poised, slender and good looking; but she had only a technical school education and no experience of life outside work and marriage at the factory in Sverdlovsk. She spoke only Russian; she had only a few dresses, plus the fine one given to her by the Chinese ambassador in Moscow. One of her many anxieties was whether the Chiang family would accept her as the wife of their first-born son. Seeing this nervousness, Madame Soong did her best to put Faina at ease. When they parted, she gave CCK an envelope of money to buy new clothes for the family. "When he saw the little Alan with large eyes and curly hair, CKS was very happy. He was delighted that this foreign daughter-in-law had provided him with a grandson. Faina felt the unusual status and identity of her father-in-law," is how Wang Mei-yu describes this first encounter (*Biography of Chiang Fang-liang* page 16).

CCK, his mother (Mao Fu-mei) and Faina in Xikou. Madame Mao is holding their first son, Hsiao-wen.

One would expect that, after being separated from his son for 12 years, CKS would welcome him to live at his home in Nanjing. But not so; instead, he sent the family to the ancestral home in Xikou, Zhejiang province, with his first wife Mao Fu-mei, the mother of CCK. CKS felt that both son and daughter-in-law needed an extended period of acclimatisation before they could enter the dangerous and demanding world of Chinese public life. The Japanese military had occupied Manchuria and large areas of northern China; its forces were close to Beijing and threatening to attack it. If they did, would a foreign power come to the aid of China? The outlook was dark and full of foreboding.

CCK was close to his mother. During his childhood, she was the person who had brought him up and he had seen little of his father. Their reunion was very emotional; they embraced and wept profusely. The first item on the agenda for the couple was learning Chinese. During his 12 years away, CCK had used Russian far more than Chinese; he had forgotten most of the characters. For Faina, it was starting from scratch. Everyone around her, including her husband, spoke with a strong local Ningbo accent, so that is the accent she learnt. Her husband gave her a Chinese name Chiang Fang-liang (蔣方良); this is how she became known in the Chinese world for the rest of her life. She had to adjust to a completely new life, in a house where she was not the mistress; she had to follow the rules of Madame Mao and learn how to be a Chinese wife.

She also had to understand the mysteries of the Chiang family, including the relations between the two wives of her father-in-law. Another challenge was how to conduct herself during traditional ceremonies and rituals, like sweeping the graves of the ancestors during the Qing Ming Festival (清明節). There was so much to learn. CKS told his son to write an account of his years in the Soviet Union; he wanted to know what he had done and to what extent he had been converted to Communism. Since he could not write it in Chinese, CCK wrote the book in Russian and a scholar translated it into Chinese; we quoted from it in Chapter One.

At the beginning, Faina could communicate with Madame Mao in sign language only. Fortunately, the two women got on very well. Probably this was because they were similar – traditional wives and mothers who accepted their station in life and the decisions of their husbands.

Faina had a new status as the daughter-in-law of China's most powerful person and the servants that came with it; but she insisted on doing housework and making food, as she had done in Sverdlovsk. She learnt the cuisine of Ningbo. She rode a bicycle around Xikou and bought food and daily necessities. She helped her mother-in-law with the housework; she hung her washing on a line outside the house. Neighbours who had never seen a blonde European gazed at her in astonishment.

She also brought strange habits the ladies of the town found it hard to accept. She rode horses and put on a bathing suit to swim in the local creek. When the neighbours complained to Madame Mao that this was improper behaviour for a lady, she explained that western women did these things. Faina continued to swim even after she became pregnant. One shocked onlooker said: "Be careful, in your belly is a little Ching-kuo." Pointing to her stomach, Faina replied: "No, it is a little Faina!" Madame Mao warmed to her. She organised a Chinese-style marriage ceremony for the couple, to complement the simple one in Sverdlovsk, and gave Faina a long Chinese gown for the occasion.

Her husband was busy writing his Soviet memoirs and practising Chinese characters; in addition, he had to read the Chinese classics his father had ordered him to study. In their free time, the couple went with Madame Mao to her favourite temple; she was a devout Buddhist. They also toured other places in the town and called on Marshal Zhang Xue-liang (張學良); he was under house arrest in a western-style villa nearby that had been built for him. He and CCK became lifelong friends.

Faina and CCK stayed in Xikou for eight months, until the end of 1937. That period was a military disaster for China. On July 7, Japan launched its all-out war. It quickly occupied Beijing and, after a devastating three-month battle, Shanghai; during the battle, the Chinese suffered 250,000 casualties and the Japanese 40,000. It was one of the biggest campaigns of World War Two, often called "Stalingrad of the Yangtze". The foreign residents who lived in the concessions in Shanghai – not touched by the battle – looked on with astonishment at the air, naval and artillery strength of the Japanese and the heroism of the Chinese soldiers fighting them; but their countries did not intervene on China's side, as CKS had hoped. On August 21, China and the Soviet Union signed a non-aggression pact. Over the next eight years, the Soviet Union provided 1,000 aircraft, 2,000 pilots and 500 military advisers to China, at a total cost of US$250 million.

Until the Japanese attack on Pearl Harbour in December 1941, no Western country helped China. The Red Army officers who had feted CCK on his final days in Moscow had kept their promises. His 12 years there, his meetings with Stalin, his fluency in Russian and his pro-Soviet behaviour played a small part in the decision to provide this invaluable aid. After conquering Shanghai, the Japanese army occupied Hangzhou and then the capital Nanjing, where they raped tens of thousands of women and killed thousands of civilians. CKS moved his government to Chongqing on the upper reaches of the

Yangtze, deep in the interior and outside the reach of the Japanese land forces.

The best of times, the worst of times

CKS decided that, after eight months in Xikou, his son had been sufficiently re-Sinicised and was ready to enter public life. It was a critical moment in China's history. The country was fighting one of the most powerful military powers in the world with the help of no-one but the Soviet Union. In assigning his son to a public position, CKS had to choose a part of China not occupied by the Japanese; he picked Nanchang (南昌), capital of Jiangxi (江西) province in southeast China. It was an administrative and commercial city, with little industry outside food processing. Ten of thousands of people seeking to escape the Japanese occupation had fled there; it was crowded and chaotic. Faina and her husband moved there, taking with them Alan and their new daughter, Chiang Hsiao-chang (蔣孝章), born in February 1937; her English name was Amy. They set up home in a small but comfortable house. CCK was appointed deputy director of the Provincial Peace Preservation Corps (江西省保安處副處長) in Nanchang.

The Japanese army continued its rapid progress across China. In October 1938, it occupied Wuhan and Guangzhou. In March 1939, a

Faina and two babies at home (Courtesy of "Academia Historica")

Japanese division reached the outskirts of Nanchang (南昌); most of the residents fled. On March 27, the Chiangs rode in an army truck south; the road was lined with hundreds of fleeing families. They went to Gannan (贛南), a remote city in the south of the province and moved into a western-style house on a hill overlooking the town. CCK was made commissioner of the fourth administrative region of Gannan (south Ganzhou) (贛南第四行政區專員), an area covering two million people in 11 counties. It was a poor area, with primitive agriculture, mining and a handful of factories; it was of no strategic interest to the Japanese. The streets were full of refugees and their children who had fled the war.

The family lived for five years in Gannan. It was the longest they stayed in one place in the mainland. Their third son Chiang Hsiao-yong (蔣孝勇) said that, during Faina's long life in the Chiang family, this period was the richest and most colourful for her. "This was because, at that time, CCK was young and full of energy. He had just started his political career and threw all his energy into his job. Because he did not occupy a major post, he very much encouraged Mother to join his work. Together they developed their life in Gannan. She not only helped him to entertain visitors to their home. She often accompanied him in his work to all parts of Gannan. She was just as she had been in the Soviet Union, an extremely active young wife." (Wang Mei-yu biography, page 24)

This meant both within and outside the house. At home, she made food and helped to entertain the guests. Outside, she helped her husband raise funds for the war effort. The two walked through the streets of the towns; the district chief and his blonde wife caught the eyes of passers-by and encouraged them to make contributions. She also visited the homes of wealthy businessmen and, in her thickly accented Mandarin, invited them to give money. She met ordinary Chinese and shared with them the pain and suffering of the anti-Japanese war.

Moved in part by the revolutionary fervour he had seen in the Soviet Union, CCK wanted to be close to his people and not a remote official, as had been the custom for centuries in China. So, he invited young people to come to his home; here Faina had an important role, in receiving the guests. The visitors asked many questions about the life of the two in the Soviet Union and how they came to marry. In this remote corner of China, foreigners were rare, especially female ones, and the object of great curiosity. CCK saw this as a way for his wife to join the society and improve her Chinese.

He encouraged the establishment of nurseries, to reduce the heavy burden of women as mothers, keepers of the house and workers in the field. When a new one was set up, Faina visited it and brought gifts and sweets. He appointed her head of one he set up in Gannan. She learnt to play mah-jong, the most popular household game in China; it became a habit for much of her life. Unfortunately, CCK launched a campaign against gambling in Gannan; wealthy women bet substantial amounts during mah-jong games in the privacy of their spacious homes. So CCK banned Faina from playing it in Gannan – although she could play it elsewhere, on family visits to Xikou and Chongqing. As a dutiful daughter-in-law, she visited these two places, to see her two mothers-in-law. Late in 1939, two Japanese bombers attacked Xikou and targeted the Chiang family home with a missile. Mao Fu-mei, CCK's mother, was killed instantly. It was a vicious attack on an innocent housewife and grandmother.

During this period in Gannan, Faina was involved in her husband's work and mission in a way she would never be again, in the mainland or in Taiwan. In his first job in the government, he needed her active help and support. Young and full of energy, the two were making their new lives together. Faina met many people – both wives of important people in Gannan and Chongqing and the ordinary citizens of Jiangxi. Her extrovert nature enabled her to adapt to a world so different from that she had known in the Soviet Union. Later, in Taiwan, CCK built a wall between his public and private spheres; he asked Faina to stay out of his public world and not ask about it. As a loyal and devoted wife, she accepted this role.

Faina gives food to soldiers. (Courtesy of "Academia Historica")

While the Japanese military was not far away, it made no attempt to conquer this poor and remote area. So, while a terrible war raged in other parts of China, Gannan was comparatively peaceful, enabling the Chiang family to lead a relatively normal life. Faina accompanied her husband on regular flights to Chongqing, to see her parents-in-law. In the national capital, she witnessed the dramatic scale of the Pacific War. She was reminded that her husband was the son of the president and head of China's military forces – and so was unlikely to have a job as modest as that in Gannan indefinitely.

A dark hour

The period in Gannan also contained one of the most difficult episodes during Faina's 50-year marriage. In 1940, her husband established a school to train cadres who would be loyal to him and carry out his policies. The students were young, patriotic Chinese eager to serve their country during this national crisis. Among them was Chang Ya-juo (章亞若), a woman in her 20s and daughter of an intellectual family in Nanchang. She had been married at 17 and had two children; her husband then committed suicide. An attractive and talented young woman, she greatly admired CCK. She wrote articles for a news agency that aimed to help the war effort. She sang and performed Beijing Opera; after watching her perform, CCK went backstage to congratulate her. He also wrote an article in a local newspaper praising her work and enthusiasm. The two were attracted to each other. CCK asked her to be his private secretary. She sometimes accompanied him on trips to outlying districts. She visited the family home and occasionally taught the two children; Faina knew her.

CKK and Miss Chang became intimate. In 1942, she became pregnant. To avoid public embarrassment, she was sent to Guilin 790 kilometres to the west, where her brother was working as a magistrate. CCK did not accompany her but sent money. At Chinese New Year, he went to visit her for the festival; domestic staff called her "Madame" (夫人). On May 21, 1942, she gave birth to twin boys. Delighted, CCK went to visit her and held them proudly in his hands. Then he flew to Chongqing to report the good news to his father. On his return to Guilin, he told Miss Chang that CKS had given them two names – Hsiao-yan (孝嚴, John) and Hsiao-tzu (孝慈, Winston). The two first names, Hsiao (孝) meaning filial piety, were the same as those of CCK's existing children. But CKS decided that they should carry the family name of their mother and not their father – this meant they would not be accepted into the Chiang family. It left Miss Chang not knowing what status she and her sons would have in the future.

A modern and educated woman, she was unwilling to live quietly. The city of Guilin was full of foreigners, especially Americans; many foreign organisations had relocated there to avoid the Japanese occupation. She began to study English; some believe that she was considering going abroad to raise her sons. But, five months later, in November, she contracted a severe stomach illness; a friend rushed her to the hospital where she stayed overnight. The next morning, to ease the stomach pains, a doctor gave her an injection. A few minutes later, she was dead; she was just 29 years old. She was buried the next day on a hillside in Guilin. Friends gave the twins to her mother to raise. Since then, many theories have circulated as to why and how she died; one is that someone decided to assassinate her – probably by poisoning – in order to prevent her becoming a public embarrassment to the Chiang family. None of these theories have been proved or accepted by historians. When he was informed, CCK was deeply upset; he wore dark glasses to hide his eyes swollen by tears. A year later, on a visit to Guilin, one of his staff told him that the burial site of Miss Chang was nearby. "Do not raise this sad matter again," he said.

How much did Faina know of all this? According to one rumour, she demanded a divorce and a return to the Soviet Union – but the Chiang family insisted on keeping the two children, so she did not go. Chiang Hsiao-yong denied this. "My mother was not that kind of person. She was able to accept a grievance. At that time, she knew of the affair between CCK and Chang Ya-juo but did not complain about it to other people. Nor did she demand a divorce. The feelings between my Father and Mother were very good. Perhaps because of this affair, Father felt guilty toward her. So, in front of their children, he was very affectionate toward her, especially in their later years." (*Biography of Chiang Fang-liang* by Wang Mei-yu, page 31 and 32). When her husband returned from trips, she greeted him at the airport with a hug and kiss – which a Chinese wife would not do in public. The two usually spoke in Russian; CCK liked to read Russian literature. Chiang Hsiao-yong said that, while his mother knew of the affair, she perhaps did not know of the twins. "She was a very intelligent woman, with sharp eyes, and saw many things. Her character was to internalise things and not say them. CCK was her world." (*Biography of Chiang Fang-liang* by Wang Mei-yu, page 33). To protect her feelings, during his lifetime, CCK never told her of this affair. It was a taboo subject in his household. The family had no contact with the twin boys and those who brought them up. If Faina looked around her, she saw the behaviour of other rich and powerful Chinese men; her father-in-law had married twice and had other relationships. In the society of that time, the wives of these men had little choice but to accept the misbehaviour of their husbands; they did not have financial independence and divorce was not socially acceptable. Their status and standard of living depended on their husband. A foreigner without her own family and support network in China, Faina was even more vulnerable.

Conversion

Faina was born into a Russian Orthodox family; it was and is the majority religion of the Russian people. But the new Soviet government aimed to eliminate religion. It confiscated church property and treated the Orthodox clergy and practitioners as "anti-revolutionary"; many were arrested, exiled and imprisoned. So, it was extremely difficult for Faina, as well as other Soviet citizens, to retain their religious faith, except in private. On her arrival in China, however, she discovered that her parents-in-law were devout Christians. This was the result of CKS's marriage to Soong Mei-ling. Soong was the fourth of six children of a wealthy businessman and former Methodist missionary. CKS first met her in 1920; he was 11 years her senior, married and a Buddhist. When he asked her parents for permission, her mother vehemently opposed the idea. First, she demanded that he divorce his wife.

For what happened then, we have the account of Pastor Chou Lian-hua (周聯華), the Baptist minister of the Chiang family for 40 years. (*Chou Lian-hua Memoirs*, page 203). "Mother said: 'We will not allow our three daughters to marry a non-Christian.' 'To marry her, I will become a Christian,' CKS replied. 'From now on, I will every day read the Bible.' Mother said: 'Mr Chiang, from the kind of man you are, if you say something, you will carry it out. If you fulfill this promise, I

Faina and Song Mei-ling (宋美齡)(Courtesy of "Academia Historica")

can answer you on the question of marriage.' After this, CKS really did read the Bible every day. In the beginning, he found it very difficult and often had to go to the house of Madame Soong to ask for help. She knew the Bible very well and constantly guided him. He could not keep up. But he made a determined effort and memorised the entire list of chapters." Finally, he won over Madame Soong and married her daughter in Shanghai on December 1, 1927. Thereafter Christianity became an integral part of his life; he read the Bible every day and attended religious services.

After CCK returned to China, his father also asked him to study the Bible and other Christian literature. This he began to do after moving to Gannan. On Easter Day 1943, in Chongqing, Pastor Pi Fan-yu (畢範宇牧師) baptised CCK and his family into the Methodist church. "CCK read the Bible and *Streams in the Desert* (荒漠甘泉) and often quoted the Bible in his speeches," wrote Pastor Chou (*Memoirs of Chou Lian-hua*, page 222).

In her later years, Christian belief became an important part of Faina's life. *Streams in the Desert* was published in 1931 by the Oriental Missionary Society in Los Angeles. The author was Lettie Cowman, an American missionary who had served in Japan. Each daily section contains a passage from the Bible and a quote from another author. Cowman wrote the book while she was watching the slow deterioration of the health of her husband Charles, after they returned to the U.S. from Japan in January 1918. It describes her hardships and her closeness with God. The title of the book comes from Isaiah 35:6: "Then will the lame leap like a deer, and the mute tongue shout for joy. Water will gush forth in the wilderness and streams in the desert." Cowman often said that it was God, not she, who wrote the book. It was particularly suitable for people like CKS and CCK; they were fighting a terrible war against an enemy with overwhelming military superiority. From where could they draw the strength to continue the battle?

The CCK family's stay in Gannan was cut short by a new Japanese offensive, Ichigo (一號作戰), launched on January 17, 1944. Within a few days, Japanese units were close to the town. CCK realised that the town was about to fall. He evacuated Faina, the two children and their maid (Ah Wang) to Chongqing, along with his key staff. He himself left on February 3. Most of the residents fled. On February 5, the Japanese army occupied the town.

Main sources for this chapter:

The Generalissimo's Son, by Jay Taylor (Harvard University Press, 2000).

Biography of Chiang Fang-liang, by Wang Mei-yu (China Times Publishing Company of Taiwan, 1997).

Memoirs of Chou Lian-hua (United Literature Publishing Company, Taipei, first edition 1994, second edition September 2016).

Chiang Fang-liang and Chiang Ching-kuo, by Zhou Yu-kou (Rye Field Publishing Company, 1993).

CHAPTER 3

Civil War and Exile, Separation and Anguish

The years between the escape from Gannan in January 1944 and the retreat to Taiwan in 1949 were turbulent and dangerous for Faina and her family, as they were for millions of Chinese. She did not know the outcome of the war with Japan; nor did she expect the Communists to win the civil war. Her husband was on the assassination list of both. During this period, she moved house several times and lived for long periods away from him. For eight months in 1949, she lived with the children in Taiwan, while he remained with his father in the mainland. She did not know if her husband would escape alive, or be captured by the Communists and imprisoned for years as a trophy of their victory.

"The years between 1944 and 1949 were a period of extreme difficulty for CCK ... in his public life, he encountered very many obstacles. Chiang Fang-liang was his greatest spiritual support." (*Biography of Chiang Fang-liang*, Wang Mei-yu page 38)

Chongqing under fire

From Gannan, Faina and the children were flown to

Chongqing, the wartime capital, where the government had relocated after the fall of Nanjing in December 1937. She had often been there before during the war to visit her parents-in-law. President Chiang chose it because it was out of the reach of the Japanese land forces; it had no railway. It had a distinctive geography, situated at the confluence of the Yangtze and Jialing rivers and surrounded by mountains.

The population swelled with the arrival of military and civil officials and of those working in the many factories moved to escape the Japanese occupation. By the end of the war, the population in the urban and suburban areas reached over a million, more than triple the 269,000 in 1932. The city's isolation, poverty and the shortages caused by the war caused conditions of life to deteriorate. Many houses were built of bamboo; people used common toilets. There was a severe housing shortage, which drove up rents.

The greatest danger came from the Japanese air force. In 1937, after the start of their all-out war on China, the Japanese military issued Article 103: "It is very important to directly strike the civilian population, to put the greatest terror into the hearts of the people of the enemy country and break their will." (《國家人文歷史》，2019 年 1 月 , *National Humanities History magazine*, January 2019).

It was the same strategy Nazi Germany would use in Britain; and, later in the war, the U.S. and Britain in Germany and Japan. The Japanese built airfields in Hubei from which the planes took off. Between February 1938 and the end of 1944, these bombings killed 32,829 civilians and soldiers in Chongqing. The Chinese air force was woefully ill-equipped to defend the city against planes that included Mitsubishi G3M bombers and Zeros, the most advanced fighter plane in the world when it was introduced in 1940. At the end of 1936, China's air force had 600 aircraft, of which 296 were fit for combat. Between 1938 and 1941, the city's air defences shot down about 100 Japanese planes.

Between February 1938 and August 1943, the Japanese air force carried out 218 raids and dropped 11,500 bombs that destroyed more than 17,600 buildings. They concentrated fire on the urban and commercial areas where most of the population lived. The Chiang family and other senior civilian and military leaders lived not in the urban area but in the outskirts, where the mountains gave them a measure of protection. The nearest the Japanese came to killing CKS was on August 30, 1940, when bombs went off a few metres from his residence. He and his family survived.

Without an air force to resist effectively this bombing, Chongqing city embarked on a vast programme of building tunnels for residents to take shelter during the raids. Short of equipment and money, the city had to rely on the efforts of thousands of people using pickaxes, shovels and other hand tools. By the start of 1941, they had built 1,700 tunnels able to shelter 460,000 people, almost equal to the urban population; the biggest was 2,500 metres long. A parallel city grew up underground, similar to the one above. A total of 674 specially trained people worked in 147 watch posts around the city to alert people of the raids and give them time to reach the tunnels. Sometimes, however, the warning was not soon enough. At 18:00 on June 5, 1941 as everyone was sitting down to dinner, the sirens sounded and everyone raced to the tunnels. One shelter 10 metres below ground could accommodate 5,000 people; but, in the rush, double that number entered. More than 1,000 died of suffocation; it was the largest tragedy of this kind during the raids.

A tragic mistake

The man who led the attack on CKS's home was a senior Japanese air force commander named Saburo Endo（遠藤三郎）. Thirty years after the war, he made public his wartime diary, in which he said that the bombing of Chongqing was an enormous mistake. It did not achieve its objective of terrorising the population; after the raids, the Japanese pilots saw life return to normal in the urban and rural areas of the city.

"For a long time, the results of the bombing have been exaggerated. If you made a judgement, our army has already turned Chongqing into a waste city. This is a mistake, an enormous mistake." The bombings did not force the CKS government to surrender, nor turn the population against it. Endo passed his opinion to the Chief of the General Staff; he was strongly criticised for this heresy.

New life in Chongqing

Faina's life changed after the move to Chongqing. In Gannan, CCK was the commissioner in charge of 11 rural counties with two million people; emperor of his little kingdom, he could largely decide his own schedule and Faina was an important part of it. But Chongqing was the national capital and CCK the son of the president. CKS saw him as a close and trustworthy adviser and someone to whom he could entrust major responsibilities. His first assignment was to train students and young soldiers. He had little time for his wife and children, however happy he was to see them. He divided his family time from his work and social life. His extrovert and lively personality attracted women. He had an affair with the daughter of an air force general; she became pregnant and went to live in the United States. (*The Generalissimo's Son,* by Jay Taylor, Chinese version, page 121) On April 25, 1945, Faina gave birth to their second son, Chiang Hsiao-wu (蔣

孝武), in Chongqing.

CCK nearly became the governor of Xinjiang (新 疆). In April 1945, his father sent him to Urumqi (烏魯木齊) to manage a complex political situation that involved local armies, the "East Turkestan Republic" (ETR), warring ethnic groups and strong Soviet influence. The ETR was a self-proclaimed state set up by Uighurs who wanted independence from the central government's control; they declared it in Kashgar on November 12, 1933. They never consolidated control over the region nor won recognition from foreign countries.

Speaking Russian and understanding well the Soviet Union, CCK would have been a good choice for governor. But CKS decided that, as the end of the war approached, his son had more important duties. In June and December 1945, CCK made two visits to Moscow as part of Chinese delegations, to discuss the post-war settlement.

During the June visit, CCK had a private meeting with Stalin, who asked after Faina and Alan, their first son who had been born in the Soviet Union. Stalin gave the present of a rifle for the young man. On January 1946, after CCK's return to Chongqing, Faina met him at the airport.

Later in 1946, CCK moved with the government back to the pre-war

Faina and three of her children (Courtesy of "Academia Historica")

capital of Nanjing, where he worked in the Ministry of Defence and was in charge of demobilising soldiers of the Youth Army (青年軍). He lived and worked out of two rooms on Zhongshan Road (中山路); it was in the centre of the city, close to major government buildings.

CCK sent Faina and the children to live in Hangzhou, capital of Zhejiang province and one of the most pleasant cities in China; it had been left almost untouched during the war. The family lived in a two-storey house close to the West Lake (西 湖), the most desirable district. Hangzhou was away from the fevered political atmosphere of the capital; and from officials seeking favours from the Chiang family by doing things for Faina and the children. Another reason to choose Hangzhou was that CCK was frequently out of the capital on work visits.

Nanjing also had a Soviet embassy. If Faina lived in the capital, she might have contact with the embassy staff; this would have aroused suspicion and rumours that could be used in the intense faction battles within the KMT. It was safer to keep her away from the embassy. In Hangzhou, she and the children could have a normal life, away from politics; the two elder children attended school. Husband and wife regularly visited each other; in Nanjing, Faina also went to see Soong Mei-ling; China's first lady gave her money, to supplement the limited funds she received from her husband, although the mother-in-law told Faina not to tell CCK about this. Madame Soong and CKS were delighted to see their Eurasian grandchildren. On one occasion, the brother of Chang Ya-juo brought the twins to Nanjing to see CCK. After this meeting, he made a promise that he would never see the two again; he kept this pledge for the rest of his life – a sign of loyalty to Faina and his family with her.

Faina knew no-one in Hangzhou. So, socially, she had to start again as she had in Gannan. One advantage was that her Ningbo accent was more easily understood than in Gannan; Ningbo is in the province of Zhejiang, of which Hangzhou is capital, so the accents are similar. She made friends with the wives of local Chinese officials. She took up Chinese painting and played mah-jong.

Move to Shanghai

In the summer of 1948, CCK was given a new and extremely difficult assignment in Shanghai, the country's industrial and commercial capital. His father sent him there to control the rampant inflation that had resulted from the civil war and widespread speculation.

On August 18, the government issued an order to its citizens to hand over their gold and silver bullion, plus all of their old currency, in

Faina and CCK with two other children

exchange for a new money called the gold yuan (金圓券). It also banned wage and price increases, strikes and demonstrations. CCK was given the task of enforcing these draconian regulations; most people said that they were unenforceable.

He moved into one of eight Spanish-style villas on 1610 Linsen Road (林森路) – now Huaihai Road (淮海路). It was a two-storey house built in 1942 of brick and wood. The sitting room was on the first floor; the bedroom, reading room and CCK's office were on the second floor. Behind was a small garden planted with camphor laurel trees and covered with Chinese parasols. A small room at the side was occupied by the police who guarded the family. CCK rose every morning at 06:00 and went jogging along Linsen Road; then he had a breakfast of congee. Faina and the children came to visit but continued to live in Hangzhou. Their third son, Chiang Hsiao-yong (蔣孝勇), was born in Shanghai in October 1948. The two elder children continued their studies in Hangzhou.

But the challenge facing CCK was too difficult. He was fighting rich and well-established families and businesses, some connected to his own family; the People's Liberation Army was winning the civil war. On November 1, CCK released a statement to the people of Shanghai, apologising for his failure to curb inflation and stabilise the economy. On November 5, he gathered his closest colleagues and told them:

"Now we have failed. I do not know where we should go and what we should do." CCK left Shanghai and rejoined his father in Nanjing.

By now CKS realised that he would lose the civil war. He could have decided to resign as president and go into exile abroad. Many members of the Kuomintang, especially the rich and well-connected, chose to leave and settle in the United States, where they had assets and friends. This would have been an easy option for CKS; his wife was a national personality in the U.S. and her family one of the wealthiest in China. The U.S. had given asylum to many members of the royal families and governments of Europe who had been overthrown by Communist revolutions.

The Chinese civil war was one of bloodiest conflicts in human history. Between 1946 and 1949, it caused an estimated six million casualties, including civilians; to this can be added hundreds of thousands of casualties between 1927 and 1937. Defeat in such a Biblical conflict would have persuaded most men that history had abandoned them. The White Russian leaders in the Russian civil war, which claimed seven-to-12 million casualties between 1917 and 1922, took the decision to leave; those who could escape went into exile in Europe and the Americas.

But CKS believed that, however badly the war was turning against him, he remained the legitimate ruler of China and that those fighting him were "Communist bandits" (共匪); he would use this term for the next 30 years. He refused to go into exile. He moved his government, military and supporters to Taiwan, an island around 180 kilometres from the mainland. It was the safest sanctuary in China and offered the protective barrier of the Taiwan Strait. From there, he would plan an invasion to recapture the mainland. At that time the PLA had almost no aircraft; its first air unit was only formed in July 1949 in Beijing, with 10 planes. So a PLA attack on Taiwan would take months, if not years, to organise. To finance his new government, CKS moved China's gold, silver and foreign currency to Taiwan in the early months of 1949; these were operations carried out at night by trusted members of the military in total secrecy. In addition, he transported to Taiwan thousands of crates of art treasures from the Palace Museum in Beijing; this was to show that he was the true "custodian" of the national culture.

Months of anxiety

The months following her husband's departure from Shanghai in November 1948 were very difficult for Faina. She lived with the children and their staff in Hangzhou. She saw little of her husband, who had become the closest confidant of his father. Both were on a death list issued by Mao Zedong who described them as "enemies of the people" (人民公敵).

There were many Communist spies within the government and military of the Nationalists; the assassination or capture of CCK would be a trophy beyond value. So his movements and plans became top secret. Faina often did not know where he was or what he was doing. He considered sending his wife and children to safety in Hong Kong or England. But he had no money to support them in the way he wished; and he refused to accept funds from the Soong family. In any case, such a move would have been politically impossible. If the son of CKS did not send his family to Taiwan, how could ordinary members of the Nationalist Party have confidence in him?

At her home in Hangzhou, Faina received a message from her husband telling her to be prepared to leave at short notice with the children. On January 21 1949, CKS announced that he was stepping down as president. That day he and CCK left Nanjing aboard the "Meiling" (美齡號) aeroplane, the property of Soong Mei-ling (SML), and flew to Hangzhou; there they picked up Faina and the children. From there, they flew to the ancestral town of Xikou. While the civil war continued to rage and the PLA won one victory after another, the Chiang family stayed for three months in Xikou.

Like many Chinese, Faina could not understand how her husband's government had lost the civil war. Now she had to accept the reality that she and her children would soon be leaving their home and moving somewhere unknown. She had been in the mainland for 12 years; with enormous effort, she had mastered the Chinese language, with a Ningbo accent, and learnt the country's culture and traditions. She had made friends and contacts. She had lived in more places than she had in the Soviet Union. She had been accepted and loved as a member of China's first family. But now, as in 1937, she would have to move again to a new place of which she knew nothing and where she had no friends. Taiwan had a tropical climate, with heavy rainfall and humidity; it was regularly hit by typhoons and earthquakes. She had no idea of the life there that awaited her and her children.

On April 23, the PLA captured the Nationalist capital of Nanjing. The next day, CCK said goodbye to Faina and the four children at Xikou. They embraced their father and were taken by car to a military airport in Ningbo; CCK did not go to see them off. With the nanny Ah Wang (阿汪姐), Faina and the four children boarded a military plane which flew them to Taichung (臺中) in central Taiwan.

Her husband stayed behind. He spent the next eight months in the mainland. The territory controlled by the Nationalist government shrank each month, so the danger to him increased. The capture, or assassination, of CCK would be a triumph for the advancing Communist armies, a scalp second only to capturing his father. Faina begged him to stay with the family in Taiwan and not put his life at risk any further. But her appeals fell on deaf ears. He considered it his duty to stand with his father, who insisted on continuing the fight in the mainland and winning more time to evacuate soldiers and civilians to Taiwan. Faina had little or no idea where her husband was; his itinerary was a closely guarded military secret. The KMT government and military was full of Communist spies eager to know where CCK and CKS were going and the place they would choose as the final base of their government.

The family went to live in a house in central Taipei, which CCK had rented for them. It was 18 Changan East Road (長安東路 18 號). The two banks that owned it offered to sell it to him, as a leader, at a "friendship" price; he refused. Faina moved to the house with the children but without her husband. In 1937, on her first journey into the unknown, he was at her side – but not this time. When she asked when he would join them, he could give no promises. "My children are very dear to me but there is a conflict between public duty and my private wish," he told her. "I have to sacrifice my private interest for the public good." (*Biography of Chiang Fang-liang* by Wang Mei-yu, page 41). So she had to face an uncertain future in a new place on her own and without her husband.

The months that followed were a nightmare. She knew the civil war was lost and that the Nationalists were moving their government and military to Taiwan. The Communist Party intended to "liberate" Taiwan; U.S. support for CKS was wavering. She wrote many letters to her husband to express her love and concern and ask him to join the family in Taipei. But he saw his duty at the side of his father. CKS did not want to desert the thousands of his soldiers trapped in different cities on the mainland; and he wanted to buy time to prepare his new base in Taiwan. At the same time, he wanted to confuse the Communists as to his final destination. He and his son's movements were a closely guarded military secret. So Faina could do nothing but wait nervously for news; her husband and father-in-law could be killed in battle or by an assassin's bullet, shot down in an airplane or captured. Each week that passed made such an outcome more likely.

By November, nearly two million civilians and military personnel, including their families, had moved to Taiwan. The final stop in the mainland for CCK and his father was the wartime capital Chongqing, where they arrived on November 14. The military situation was hopeless. CCK urged his father to leave before it was too late. In early December, they flew to Chengdu. PLA forces under Deng Xiaoping (鄧小平) captured Chongqing and advanced on Chengdu. Deng was the fellow student CCK had met in Moscow in 1926; as the two walked along the banks of the Moscow River and passionately debated the future of China, they could not have imagined this tragic and blood-soaked finale.

Finally, on December 16, father and son left Chengdu in a DC-4 transport plane for the perilous 1,824-kilometre journey to Taipei. For nearly all the journey, the plane flew over areas controlled by the new Communist government. They completed the journey only due to the bravery and technical brilliance of pilot and navigator, Major Konsin Shah Kung-chuan (夏功權). Clouds prevented a view of the terrain below; there were no radio stations to provide guidance. Shah used his experience and instinct to guide the plane and landed it in Taipei at 2130. Only then could Faina and the children finally breathe again.

Main sources for this chapter:

The Generalissimo's Son, by Jay Taylor (Harvard University Press 2000).

Biography of Chiang Fang-liang, by Wang Mei-yu (China Times Publishing Company of Taiwan, 1997).

Chongqing in wartime: building the world's biggest underground city. Article in *National Humanities History magazine*, January 2019. This gives an excellent account of the bombing campaign.

Chiang Fang-liang and Chiang Ching-kuo, by Zhou Yu-kou (Rye Field Publishing Company, 1993).

China Daily 12/1/2015 article on CCK's home in Shanghai in 1948.

CHAPTER 4

Normal Life – the 1950s

The new family home in Taipei was in Changan Dong Road (長安東路), close to Zhongshan North Road (中山北路), the city's main north-south artery. It was an old-fashioned Japanese house; the family furnished it in a modest style. They lived in it for 18 years, until 1967, when it was demolished as part of a road-expansion project.

It was the centre of Faina's universe; she was a homemaker who devoted her energy to her family and children. Because of CCK's status, the family was given domestic staff. But Faina insisted on doing many of the household chores herself; she was house-proud, cleaning the curtains and the sofas. She made food, especially the dishes that her four children liked. She raised chickens in the back garden, to provide eggs for the family. The house was close to a railway line. This made the windows dirty with dust and soot – neighbours could see Faina cleaning the windows.

The house was within walking distance of downtown Taipei. This meant that the family could have something of a "normal" life – as normal, that is, as it could be for the second most powerful person in the land. When

Faina plants a tree. (Courtesy of "Academia Historica")

CCK was not busy, he and Faina walked with the children on nearby streets, full of shops, restaurants and small food stalls. Both loved American films. Sometimes, after dinner, they took the jeep out of the garage and went to a local cinema. They queued to buy tickets and sat down in the middle of the audience.

They went to local restaurants, sometimes without bodyguards. While CCK had a driver assigned to him, he often liked to drive his car. "His driver was constantly complaining that he could not find him," said Chiang family bodyguard Weng Yuan in his autobiography (*My Years at the Side of Chiang Kai-shek and his Son*, page 263). CCK did not enjoy the constant surveillance and sometimes wished to meet people privately, without others knowing.

The couple enjoyed performances of Beijing Opera, for which Faina had developed a taste in the mainland. Faina enjoyed this proximity to ordinary people and ordinary life. In the evening, the couple entertained friends and their wives at home or at a guest house in Yangmingshan (陽明山), a mountain area overlooking the city to the north. CCK often wore a long Chinese gown and served the dishes. At the arrival of each dish, the guests toasted it with rice liquor. Most became tipsy and some fell to the floor; after years of training in the Soviet Union, CCK could hold his liquor as well as anyone. Taiwan could not import liquor like vodka directly from the Soviet Union; but friends ordered large quantities from Hong Kong, so that Faina could enjoy it with her husband. She could hold her liquor as well as, or better than, he.

In the late 1950s, CCK and Faina would also attend staff parties at the club of the U.S. in Taipei where they had to wear the designated costume, such as cowboy shirts.

She did volunteer work to help local schools. Her small circle of friends embraced Chinese and foreigners, including White Russians. She liked smoking; she stored packets in the clothes cupboards, to hide them from her children. But they discovered the hiding places and took packets to school. It was an era when few Chinese women smoked; it was better not to advertise the fact.

After their arrival in Taiwan, Faina and her husband put much energy into learning English. The United States was the principal ally of the Republic of China (Taiwan's official name), so they met many Americans for work and social activities. CCK hired the wife of the military attaché at the U.S. embassy to teach Faina. Both reached a good standard. In Taiwan, it became Faina's most important means of communication with the outside world; she followed domestic and foreign news through the English media.

A keen sportswoman in Sverdlovsk, Faina took up ten-pin bowling at a nearby centre. According to Chiang family bodyguard Weng Yuan (翁元), she excelled at bowling but, after a short time, stopped going for reasons that were not clear (*My Years at the Side of CKS and His Son*, by Weng Yuan, page 225). Then she took up golf and went with friends to the Old Tamsui course (老淡水高爾夫球場), north of the city; it is the oldest and most prestigious course in Taiwan, founded by Japanese in June 1919. It is in a hilly area near the mouth of the Tamsui River and winds from the sea make it a challenging course. Her husband encouraged her to play.

"But, after several weeks, she also stopped playing. Probably it was because of her asthma (氣喘) and the difficulty of finding partners. So she stored her golf clubs away on a high shelf and lost interest in the sport." The asthma was due in part to her smoking; later doctors advised her to give up smoking because of the asthma and other lung conditions.

CCK rose at dawn; he did exercise and took a shower. At 06:30, his staff brought him files to read. He went to the kitchen and had a breakfast of congee with pickled vegetables, and sometimes poached eggs. Once he went jogging on Yangmingshan, he got lost because he was not familiar with his new environment. Alarmed, his father had to send staff to look for him. After that, he was accompanied by bodyguards. In the early years in Taipei, CCK went jogging through the city streets; but his security staff told him it was too dangerous. So his main exercise became long walks in the countryside and on Yangmingshan.

CKS did not like his son living in Changan Dong Road; he considered it not secure enough from possible attack. Assassins could be sent by the rival party or Taiwan people angry at the martial law that CKS imposed when he moved to the island. The KMT dealt ruthlessly with anyone suspected of supporting Communism or independence for Taiwan. According to official figures, the security forces arrested 29,407 people under martial law between 1949 and 1987, when it was lifted; about 15 per cent of these – 4,500 – were executed. In the mid-1950s, there were more than 10,000 political prisoners on Green Island (綠島), offshore the southeast of Taiwan.

In the 1950s, CCK played a key role in the administration of Taiwan's military and internal security institutions. In 1951, he set up a Political Staff College (政治作戰學校) in Peitou (北投), north of Taipei, to train political officers for the armed forces. He held senior posts in intelligence and sent his staff to the Taiwan Garrison Command (臺灣警備總司令部), the agency in charge of enforcing martial law. So he was one of the authors of the "white terror" which the government considered necessary to consolidate its rule in these early years in Taiwan.

Faina knew little of her husband's work. Things had greatly changed since their time together in Gannan in Jiangxi in the late 1930s. There she had shared many of his professional activities and hosted people in their home. His status in the government there was modest – the chief of a rural area with a population of two million. But, after they moved to Chongqing in 1944, his position changed abruptly – he became a close advisor to his father, the President. He told Faina not to inquire about his work; she accepted this and did not ask. He did not want her to be used by those who wanted favours from him.

So Faina concentrated her energies on the home, the family and the children. One of them, Chiang Hsiao-yong, said: "As Father's responsibilities became more onerous, he asked Mother not to inquire about them. She considered this reasonable and completely obeyed him. She became a simple housewife. She became a different person from the very active one she had been in Gannan. She did not like to appear in public places nor speak to strangers. Her character became quiet and sometimes a little lonely." *(Biography of Chiang Fang-liang*, by Wang Mei-yu, page 47). In this, she was a sharp contrast to Soong Mei-ling, the most famous Chinese woman in the world. She loved the spotlight and maintained a wide network of friends and contacts, Chinese and foreign. In the West, Madame Soong had a bigger profile than her husband. She advised him on many issues, especially foreign affairs.

Another thing Faina knew little or nothing about was the fact that her husband had arranged the transfer of his twin sons by Chang Ya-juo to Taiwan. They lived with their grandmother and uncle in a small house in Hsinchu (新竹), a town 90 kilometres southwest of Taipei.

It was a different world from the gilded cage of the Chiang family in Taipei, surrounded by servants and bodyguards. Hsinchu had a mixed population, including Taiwanese and Hakka, who each spoke their own language. The twins became fluent in both languages, as well as Mandarin which the Nationalists had brought from the mainland and made the official language. The Aboriginals who lived in Taiwan before the arrival of the Han Chinese also spoke their own languages.

By contrast, their half-brothers and sister in Taipei grew up in a world dominated by mainlanders who spoke only Mandarin. CCK kept his promise to have no direct contact with the twins; the topic was taboo in the family home. He did not want to hurt the feelings of his wife who had stayed loyal to him through his affair; nor did he wish to be reminded of his own infidelity. At the same time, he sent a trusted advisor to Hsinchu to know how the family was doing. The exchanges were not smooth; Ya-juo's brother believed that her sister had been murdered and was suspicious of the government.

In the late 1950s, the two sons chose English names – John for Hsiao-yen (孝嚴) and Winston for Hsiao-tzu (孝慈). At that time, their grandmother informed them of the identity of their father; previously, they had believed that he had been left behind in the mainland and was unable to escape to Taiwan. She told them not to reveal the truth of their paternity to anyone. So they understood that the Chiang family would not acknowledge them. CCK is rumoured to have had affairs with women during the 1950s; but these also did not disturb the atmosphere in Changan Dong Lu.

Home from home – Astoria Café and Bakery

One of the favourite haunts of Faina and her husband during the 1950s was the Astoria Café and Bakery, the first Russian restaurant in Taipei. It opened for business on October 30, 1949 at 7 Wu Chang Street (武昌街) in downtown Taipei, close to the city's main railway station. The owners were six Russians who had, like the Kuomintang, escaped from the Communists in the mainland. They were among 100 White Russians from Shanghai who had come to Taiwan; it was their second exile – the first followed the Bolshevik revolution in their homeland in 1917. The largest cities of White Russian settlement in China were Harbin and Shanghai.

Among the arrivals in Taiwan was George Elsner Constantin Enobche.

Chiang Ching-kuo and Faina (far left) attend a party at the Astoria Café. (A framed photo on the café's wall shot by Mark O'Neill)

Born in 1893 into a family related to Tsar Nicholas II, he served in the guard force protecting the Tsar. During World War One, he was an officer in the Imperial Russian Army fighting the Germans: then in the White Russian forces fighting the Red Army in the civil war. After the Soviet victory, he moved to Harbin, where he lived for three years. Then he settled in the French concession of Shanghai; he worked in the Public Works Department, responsible for construction of new homes for the French military. He spoke Russian, French, English and Spanish.

The life of Elsner and the other Russians in Taiwan was difficult. They had lost their nationality and found it hard to get work; they lived in humble circumstances and had few assets. Few spoke Chinese or knew anyone in Taiwan; they needed help from local people. After 1949, the Republic of China (RoC) became a strong ally of the United States in the Cold War and fiercely anti-Communist. One national slogan was "Oppose Chinese Communists and the Soviet Union" (反 共 抗 俄); this made employers reluctant to hire Russians, even though they themselves were refugees from Communism. They also could find nowhere to eat Russian food.

Then Elsner had the good fortune to meet Archibold Chien (簡錦錐), a young Taiwanese who spoke English. The two became good friends. Born in 1932, Chien made a visit in 1940 with his elder brother to

Shanghai; he saw the Astoria Confectionery and Tea-Room, close to the Astor House Hotel, next to the Suzhou Creek. It sold buns and bread loaves, caramels, candy and French and Russian cakes; the staff were mainly Russian and Chinese. Elsner and Chien put their heads together and decided to set up a similar establishment in Taipei, also called the Astoria.

Karen Chien (簡靜惠), the daughter of Archibold Chien, remembers: "My parents set up the Astoria Café and Bakery in 1949. When the KMT government came here in 1949, about 100 White Russians came from Shanghai at the same time. They had no nationality, having fled from the Soviet Union in 1917, and could not return home. My father was very interested in art and culture and developed a strong feeling for these Russians. My parents learnt the Russian language from them, and many other things. My parents gave some of them a home in our house in Zhongshan North Road (中山北路). Russians had no income and depended on my parents, who were very kind to them. The 1950s and 1960s was a period of the campaign 'Oppose Communism and the Soviet Union' (反共抗俄). Sometimes, if anyone spoke Russian on the street, the Russians were abused. But my parents were not afraid; they had the courage to help them. My parents installed fans in the café to keep people cool during the summer. The Russians helped them in designing the cuisine. Astoria served the first chocolate cake in Taiwan. It became the Russian community for Faina after she moved here." (Note: interview with Karen Chien, Taipei, 17/12/2018)

The six shareholders were white Russians, including Elsner. Chien found the premises for the café. It was a building opposite a temple that no Taiwanese dared to rent, for fear of attracting evil spirits. The landlord was reluctant to sign a lease with Russians, such was the intense anti-Soviet atmosphere of the time. But Chien persuaded him by showing him documents that one of the six had become a naturalised Chinese and most had served in the Nationalist army – they too were refugees from Communism. Convinced, the landlord signed a lease with the six, for a monthly rent of NT$2,000 (US$65.50, using the 2019 conversion rate). Chien invested US$500 for the purchase of kitchen equipment and building materials.

The Astoria had a bakery on the ground floor and restaurant on two floors above. Elsner and Chien became close friends; the Russian became a father figure to the young Taiwanese, who was fascinated by his extraordinary history. He had started life at the opposite end of Russian society from Faina – attending a school for the nobility and going to serve in an elite unit protecting Tsar Nicholas. In his military career, he had fought the German and Austrian armies on the Eastern Front. Then he fought the Red Army, before moving to Harbin and Shanghai. He showed the young man photographs of the Tsar and his

Karen Chien (簡靜惠) and her mother Chien Hwang Bi-hsia (簡黃碧霞) in front of the Astoria Café (明星咖啡館) (Photo by Mark O'Neill)

family; he kept these and other precious souvenirs of his homeland in a small wooden box he carried throughout his long odyssey. Now in Taipei, he had no nationality, few friends and few assets – while Faina was married to the son of the president and lived in a comfortable house with servants and bodyguards.

The café became an important meeting place for the white Russians and other expatriates in Taipei. It served full Russian meals, with soups, biscuits and pastries. This included a soft candy once available only to the Tsar and his family and Mazurka walnut cakes. The Russians loved to go there – they could speak their mother tongue, eat their favourite foods, listen to Tchaikovsky and Rachmaninoff, drink wine and forget the tragedies that had driven them so far from home.

Soon after she arrived in Taipei, Faina started to go there with her husband. Like the other Russians, she was dislocated; she had spent 12 years in mainland China and knew it better than the Soviet Union. Then, suddenly, she had to move again. Taiwan had a wet, humid climate unlike any she had experienced before. Most of its people spoke Taiwanese, which was incomprehensible to her. The 1950s were a decade of tension and uncertainty. Fearing a land invasion and air attacks from the mainland, Taiwan invested heavily in its military; government institutions built tunnels and underground shelters where people could take refuge in the event of attack. The media was heavily

蔣方良女士的最愛
明星俄羅斯軟糖

西元1949年由5位白俄人與台灣簡錦錐先生合夥於臺北市武昌街開設『明星西點咖啡』，店內這道俄羅斯軟糖就是俄羅斯沙皇專屬的御用點心。
俄羅斯軟糖用手指輕輕剝開白泡泡的軟糖，ㄉㄨㄞ、ㄉㄨㄞ觸感吃起來超彈牙，甜而不膩，也是蔣經國夫人蔣方良的最愛。純手工製作不含防腐劑，冰凍過後別有一番風味，無須解凍即可食用。這就是全台絕無僅有的俄羅斯軟糖。

明星西點咖啡 台北市武昌街一段七號 www.astoria.com.tw

Notice of Faina's favourite dishes at the Astoria (Photo by Mark O'Neill) censored; people were careful what they said. Visits to the Astoria were an opportunity to put all those things out of the mind. Faina often went with her husband; there they were not called by their Chinese names but Nikolai and Faina. "The relaxed atmosphere in the café enabled her to avoid the gaze of being a member of the first family," wrote Archibold Chien in his autobiography (*Café Astoria*, page 119). "She became a regular customer. Sometimes she brought Nikolai (CCK) to take part in gatherings of Russians. Sometimes she brought the four children to eat borscht (羅宋湯) and western food … Always she had a broad smile and walked with a light and happy step."

Karen Chien said that Faina and her husband came for a regular weekly meeting of the community, to drink, eat, dance and meet the other Russians. "The two spoke both Russian and Chinese to each other. She spoke with a strong Ningbo accent. She had a good capacity for liquor, better than her husband. The Russians could not find jobs in Taiwan. Some emigrated and some passed away." (Note: interview with Karen Chien, Taipei, 17/12/2018)

The biggest day in the calendar was the Russian New Year in early January. From the early morning, chefs prepared chicken, steak, suckling pig and many kinds of cakes and drinks. Around 2200, the guests started to arrive, dressed in traditional costume; they said their prayers and extended New Year greetings to their friends. Some played musical instruments. When the clock struck midnight, everyone toasted the others and emptied their glasses. "Sometimes, Nikolai drank several glasses of vodka and danced with great fervour, to the cheers of the audience. Faina could not hold back and sang Russian folk songs. The energy and enthusiasm of the two were like that any of young couple." (*Café Astoria*, page 120). So it was for Faina's first 10 years in Taipei.

Spartan life

In the late 1950s, as CCK was taking more senior posts in the government, he instructed Faina to avoid extravagance. So she sent a letter in Russian to the Astoria: "Because of the post held by my husband, in future I will not be able to attend private parties there." After that, she and CCK never attended a New Year party there. Her appearances were restricted to occasional visits to drink coffee and eat cakes. She remained good friends with Archibold Chien and his family; they visited the homes of the other and sometimes went to the cinema together.

Her first grand-daughter, Chiang You-mei (蔣友梅), went to the same primary and junior middle school as Chien's daughter Karen (簡靜惠); the two became close friends. Karen often stayed over at Faina's house. In 1969, CCK became "Vice Premier" and the family moved to the Seven Seas villa in northern Taipei; it was a long distance from the Astoria. Faina's social life became more restricted. She did not go to the coffee house any more; but she sent her driver to buy borscht and Russian bread.

By now, many of the White Russians had left Taiwan and settled in other countries. Chien became the sole owner. In 1964, he added two Chinese characters Ming Xing (明星 , meaning "famous star") over the front of the cafe, to reflect its new character. It became a favourite place for Taiwan authors and artists; they liked its European ambience and the fact that they could sit for several hours over one cup of coffee.

A Remarkable Woman

Karen Chien retains a strong memory of Faina: "I was at primary and early middle school with Chiang You-mei, from the ages of two to 13. She was my best friend. So I often met Faina. She was easy to get along with, not like the wife of an official; she had no airs. She was very sociable.

"The house had a large kitchen; she would go in, check and taste the dishes, to ensure the quality. Her family were very kind to me. I often went to their house and slept over. It was close to our house. I called Faina 'Grandma'. She treated me like a grand-daughter; she made the food I liked, gave me chocolate and gifts on my birthday. The family included me in activities at their home and invited me to travel with them. They treated me like a member of the family. They loved me as I was.

"When I was seven years old, I saw the film *Ben Hur* at their house. I also saw a performance of Chinese Opera at the home of President Chiang Kai-shek. All these experiences made me what I am today. Until I was 13, I did not know they were the first family. For senior middle school, You-mei went to the U.K. to study and I saw less of the family after that. Personally, Faina was a loving, remarkable grandma. I truly miss her and respect her from the bottom of my heart." (Interview with Karen Chien, 17/12/2018)

Thinking of home

Most people who leave their home country and live abroad like to keep contact with the family and friends they left behind. Who knows how long the exile will last and whether it will be a happy one?

Like thousands of others who left the Soviet Union, voluntarily or involuntarily, Faina was a victim of the Cold War that followed World War Two. The RoC joined the U.S.-led alliance that opposed the Soviet Union and its allies all over the world. Fiercely anti-Communist, the RoC was against not only its enemy in mainland China but also Beijing's main backer, the Soviet Union.

Moscow and Beijing were close allies. This meant that no direct contact with anyone in the Soviet Union – or China – was allowed, however innocent. Letters to the other side were intercepted by the police in Taiwan and the Soviet Union; they wanted to know why the author had contact with anyone in the enemy country. So letters could only be sent through third countries and carried by people in whom the sender had complete confidence. Faina's sister and friends lived in Sverdlovsk, a closed military city where foreigners were not allowed; its residents had no right to contact someone abroad.

An additional obstacle for Faina was the prominent position of her husband. He had spent 12 years in the Soviet Union where he knew many Communists, both Russian and Chinese. In January 1926, Deng Xiaoping had studied in the same university class in Moscow; they had walked beside the Moscow River discussing politics. CCK had to be "whiter than white"; so his political opponents could not accuse him of being pro-Moscow or sympathetic to Communism. The many American representatives in Taiwan were also watching him closely, as the man likely to become the next successor: would he be as friendly to the U.S. as his father and step-mother? This was another reason to keep Faina out of the public eye and any political activity, so that no-one would ask why he had a Soviet wife. In an era when the government had complete control over the media, this was easy to do. People were afraid to ask sensitive questions.

Those in Taiwan with relatives and friends in the mainland – of whom there were hundreds of thousands – sent letters and parcels to them via Hong Kong or Japan; the courier had to be someone on whom they could rely to keep the delivery confidential. It was even more difficult to send anything to the Soviet Union. In the 1950s, the colonial government in Hong Kong refused Moscow's request to open a diplomatic office there. So, with the encouragement of her husband, Faina gradually lost touch with her family and friends at home. He kept her out of the public eye. Most Taiwan people knew little or nothing about her.

"I think Faina spoke good Chinese but I never heard her voice," said Lin Kuo-keung, a Taipei taxi driver. "She stayed out of the limelight. She was foreign, not Chinese; she was a housewife, mother and wife." Wang Chi-gang, a second taxi driver, said: "We knew nothing about her. The media carried nothing about her. It was martial law. Who dared to ask such questions?" Some even believed that she was a member of the Tsarist royal family banished to Siberia, where she met her husband. "She was a handsome Russian lady," said Lin. "But it was an anti-Soviet period, so her husband did not allow her to be seen in public. Many Taiwan people did not know about her. She had a Chinese name, so some people thought she was Chinese."

In the 1960s, a U.S. Congressman met Faina and asked her if she was "White Russian" or "Red Russian". She answered: "I think you could say that I was born as one and later became the other." (我想你可以說，我生下來是一種人，後來卻長成另一種。) (*The Generalissimo's Son*, by Jay Taylor, Chinese version, page 304)

If, somehow, Faina received a letter from her sister Anna or her friends in Sverdlovsk, how could she know that they had actually written it? Or was it composed by them under duress from the KGB for a political purpose? So, for at least the first 30 years after the move to Taiwan, Faina could not contemplate a visit to her home country. In November 1970, CCK received through an emissary a letter from Faina's sister Anna, whom she had not seen in 33 years. She was the person in the Soviet Union who meant the most to her. After reading it, CCK threw it into a bag for classified documents that were to be destroyed. He told his staff not to mention the letter to anyone. It seems that CCK decided that the letter would only make his wife unhappy, in thinking of her only family member and the person who had brought her up.

The destruction of the letter tells us about how CCK saw his wife. Her role had to serve his interests, including his political ambition. In Taiwan, no less than in the mainland, he would not allow her Soviet background to have a negative influence on his career; best for her was to have no contact with anyone from the Soviet Union, official or personal. Any such contact, however innocent, could be detected by Taiwan or American intelligence and considered suspect. CCK wanted Faina to become a Chinese wife and cut all ties to her motherland; he succeeded in this, and she accepted the role.

Main sources for this chapter:

The Generalissimo's Son, by Jay Taylor (Harvard University Press, 2000).

Biography of Chiang Fang-liang, by Wang Mei-yu (China Times Publishing Company of Taiwan, 1997).

My Years at the Side of Chiang Kai-shek and his Son, by Weng Yuan (China Times Publishing Company of Taiwan, 2015).

Chiang Fang-liang and Chiang Ching-kuo, by Zhou Yu-kou (Rye Field Publishing Company, 1993).

Café Astoria, by Archibold Chien Jin-tsui (INK Yinke Literature and Life Magazine Publishing Company, New Taipei City, June 2015).

Interview with Karen Chien (簡 靜 惠), daughter of Chien Jin-tsui, in Taipei, 17/12/2018.

CHAPTER 5

Running a Country is Easy, Running a Family Hard

In speaking of his children, CCK once said: "to run a country is easy, to run a family is hard" (治國易，治家難) (*The Generalissimo's Son,* by Jay Taylor, Chinese version, page 304). Faina and her husband had three sons and one daughter. They brought their parents great joy and great grief. They were companions throughout their long odyssey. The oldest, Alan or Hsiao-wen, was with them for the whole journey, from Sverdlovsk to their many homes in the mainland and then Taiwan. The other three were born in the mainland.

To be born into the "honourable family" was a blessing and a curse. It gave the children opportunities not available to others. Their name and family opened many doors; people wanted to help them, some in the hope of personal or political benefit. They had servants, bodyguards and money; ordinary people, even the police, feared to confront them, however bad their behaviour, because of their special status.

But this status was also a curse. If someone was kind to them, did they do so out of friendship or because they wanted something? Did people tell them the truth or what they wanted to hear? Could they find out what

CCK, Faina and their four children

was really going on? Their status and their bodyguards cut them off from the lives of ordinary Taiwan people. This was especially true in the early years of martial law. The government controlled everything, including the media; it only reported official news. People were often afraid to say anything other than the "official" truth. Another burden for the children was that Faina, CCK and his parents had high expectations of them – if not to become leaders, then at least to hold important posts in the military, the government or the foreign service. It was a heavy burden to carry.

Wayward son

The eldest son, Hsiao-wen (孝文), whose English name was Alan, was born in Sverdlovsk on December 14, 1935. He accompanied his parents during their most difficult period in the Soviet Union, the final six months when CCK was out of work and they were relying on Faina's salary alone. He was their companion during their darkest hours; this gave him a special place in the heart of his parents. "From a young age, we brothers could feel that, of the four children, the one my parents most loved was big brother Alan," said younger brother Eddie Chiang Hsiao-yong (*Biography of Chiang Fang-liang* by Wang Mei-yu, page 60). "Their hopes toward him were especially high. He was much older than we two younger brothers and the attitude toward him was different … the hopes and demands of Grandfather and my parents toward him were very high." Alan attended schools in the mainland cities where they lived. He was tall and handsome.

In Taiwan, he was not a good student; he lacked discipline. What he enjoyed was alcohol, parties and guns. He attended the Cheng Kung (成功中學) High School in Taipei – aptly named "success", reflecting the hopes of the parents of the students. In his biography, a bodyguard of the Chiang family, Weng Yuan (翁元), gives a vivid description of the teenager Alan (*My Years at the Side of Chiang Kai-shek and his Son*, page 239-40).

"At High School, because he did not like to study, his exam marks were poor, with many red lines. At the time, the school sent the marks to the parents by post. Alan gave us (bodyguards) the letters and told us to store them. We did not know the contents and could only follow his orders." Unfortunately for Alan, the school principal had worked with CCK in Gannan; surprised that Father did not respond to the poor grades, he sought CCK in person and explained the situation.

"When CCK found out, he was very upset. He took a stick and beat Alan, so much so that Alan ran all over the room. When Faina heard this, she could not bear it and asked him to stop. But he continued, and Faina wept. He put down the stick and ordered Alan to kneel.

CCK, Faina and Chiang Hsiao-wen (Courtesy of "Academia Historica")

Faina continued to weep and finally persuaded her husband to let Alan get up … CCK was extremely strict toward Alan. For him, he was the most fearsome father," said Weng (page 236).

Sometimes, to escape, Alan waited until his father had gone to sleep. Then he had the bodyguards push the family jeep dozens of metres down the street; if he started it in the garage, Father would hear. Then he drove off for a night of carousing with friends.

"From an early age, Alan liked guns. From his high school years, we often saw him carrying a revolver. Since this interest was encouraged by CKS, no-one dared to stop him … One day, a plain-clothes bodyguard called Li Zhi-chu (李之楚) went off duty and back to his room to rest. He found Alan lying on his bed and holding the revolver. 'Don't move,' Alan said, pointing it at him. 'Stop fooling about, you could kill someone,' Li replied. By mistake, Alan fired a shot; it entered Li's chest. He slumped on the floor in a pool of blood. Alan went white. Li was rushed to hospital and, fortunately, survived; the bullet was only a few centimetres from his heart. CCK expressed to Li his deep remorse and sent him to a new position at a cement company in Kaohsiung, at the other end of the island." (Weng Yuan autobiography, page 238)

In another incident, Alan stormed out of the house after a row, carrying his gun. He went to a nearby fruit stall and shouted at the owner; the man was terrified. His bodyguard reported the incident to CCK. After Alan returned home, his father strongly rebuked him. Then, with the bodyguard, he went in person to the stall owner and apologised (*Chiang Fang-liang and Chiang Ching-kuo by Zhou Yu-kou*, page 234).

On another occasion, Alan drove his car while drunk; there was an accident. CCK told a friend that he was even considering cutting off ties with Alan, in an attempt to make him wake up (Zhou Yu-kou book, same page). To find out what his son was up to, CCK engaged a trusted police officer to follow him and his brother Hsiao-wu (孝武, Alex) and compile a detailed report – where they went, with whom they had meals and had arguments and when they took a gun. With this report in hand, Father summoned the two boys and demanded an explanation. If they could not provide a satisfactory one, he would criticise them sharply. One member of staff described the relationship between father and the two sons as "mice facing a cat" (老鼠見到貓) – they dared not talk back to him. (Zhou Yu-kou book, page 236).

In another incident described by Weng, after his marriage, Alan had a row with his wife and stormed out. He got completely drunk in a restaurant; everyone told him not to go home in that condition. "When Faina discovered he was not at home, she sent people everywhere to search for him. The bodyguards tracked him down at the Jiaxin Building (嘉新大樓) and advised him to go home. He still refused." (Weng Yuan autobiography, pages 242-243)

The two sons were fearful of not living up to the high expectations of their father. "For the sake of Alan's health, CCK decided to advise him in person to give up drinking. To set an example, he said that he would give up and hoped that this would move his son to follow him," said Eddie (Faina biography, Wang Mei-yu, page 68). "For the next 20-30 years, Father did not touch a drop. But, sadly, it did not improve Alan's health. He continued to drink heavily, as before."

Trapped in the Middle

Faina was trapped in the middle of this animosity between her husband and her two elder sons. She doted on her children and found her husband's harsh treatment hard to bear. But she did not want to argue with him: nor did she want to complain about him to those outside the family. So she could only internalise her sadness and disappointment. CCK was out of the house most of the time; as he climbed the political ladder, his family time diminished. His children did not have the same help from their father with homework and personal matters as their classmates. Faina did her best. She diligently studied Chinese and English; but her level was not good enough to give the help they required. In addition, she had a limited understanding of the society outside the home and of the people associating with her sons.

In the 1950s and 60s, Taiwan was a complicated society. The mainlanders who came with CKS from the mainland in 1949 were a minority. But they dominated the government, the army and the security services; the native Taiwanese were strong in agriculture, commerce and industry. Relations between the two groups were tense after the failed uprising in 1947 and the thousands of arrests and executions that followed. The island was under martial law; the legal basis for this was the "Temporary Provisions Effective During the Period of Communist Rebellion" passed by the Assembly in April 1948. Under martial law, the institution responsible for internal security was the Taiwan Garrison Command (TGC, 臺灣警備總司令部). It had sweeping powers of arrest, detention and interrogation. Its agents in plain clothes arrested suspects at their homes in the middle of the night and took them away; the families had no legal recourse. The TGC was hated by the majority of people, especially native Taiwanese, who saw it as a tool of dictatorship that operated outside the law.

In August 1992, "President" Lee Teng-hui abolished the TGC; it was one of the most popular measures he took in his tenure. During martial law in the 1950s and 1960s, power was highly centralised and secretive. The media only published what the government told it to. In this atmosphere, many people sought favours from the Chiang family; some saw friendship with Alan and Alex as a way to win these favours – a good reason to invite them out for a night of pleasure and give bottles of whisky and brandy. When he was in trouble, Alan sometimes called the local TGC commander for help; few people in Taiwan dared to do that.

Another cause of conflict was language. When the KMT took control of Taiwan in 1945, they made Mandarin the compulsory language, in schools, the media, the government and public life. But the native tongue of local people was Taiwanese; many also spoke Japanese, after 50 years of colonial rule when it had been the language of education, government and business. But neither language had any legal status. Children who spoke Taiwanese in the school playground were chastised by the teachers; some had to wear a placard saying "I must speak Mandarin". Worst off were those unable to speak or read Mandarin, especially the elderly.

In another sign of resistance to Nationalist policies, people in the 1980s discreetly installed satellite dishes so that they could watch television programmes from Japan, especially Sumo wrestling and drama series. Japanese teachers who had worked in schools in Taiwan before 1945 were invited back by their students for joyful reunions and haiku competitions; many young people went to study in universities in Japan. No country in Asia made Japanese more welcome nor looked back to its colonial period with more nostalgia.

How could Faina follow all these complex relationships and know who was related to whom? Her husband could bury himself in his work and political career outside the home; but Faina's world was a smaller one and the family was its centre. She was left at home wondering why her sons had turned out as they did.

CCK could not tolerate the drinking bouts and rowdy behaviour of the two sons, especially because they reflected badly on the Chiang family and fuelled public resentment against the privileges they enjoyed. So he criticised them fiercely. "No matter how he hit them, it did not change the sons' behaviour. This strict approach by CCK was not effective, even less so for Faina. Each time she found herself between her husband and his sons; she spoke kind words to calm the situation. At that moment, her sons listened to what she said. But, a little later, CCK would lose his temper, leaving her in a difficult position. She was a simple housewife but the Chiang family was anything but simple. Her husband was busy with public affairs far from her that she could not ask about. Her sons were always doing things to worry her. So her role as mother was very difficult." (Faina biography, Wang Mei-yu, page 78)

Military Academy

In 1955, in accord with his father's wishes, Alan went to the Military Academy (陸軍官校) in Fengshan, Kaohsiung (高雄鳳山). But, in his third year, he said that he could no longer endure the pressure of being the grandson of CKS. His parents decided to send him to the United States, a country where he could be a normal person, without the privileges he enjoyed at home. CCK used his American connections to secure him a place at the Virginia Military Institute (VMI), the oldest state-supported military academy in the U.S. founded in 1839.

As of 2019 VMI had 1,700 cadets, of whom nearly 90 per cent were men. According to its website, it "offers a rigorous education that includes a broad undergraduate program with majors in engineering, science, liberal arts, and social sciences. Woven into every curriculum is leadership and character development that benefit graduates for life. The combination of military and academic training constitutes a strenuous program requiring diligent application and conscientious attention to both academic and military duties. Barracks rooms are furnished sparingly and three, four, five or six cadets share a room. They have equal responsibility for keeping the room clean and in order for daily inspection."

But Alan's English was not up to the required standard. He could not adapt to the new environment; he dropped out after eight weeks. He moved to a small school in Kansas, for a short time, and then Armstrong College, a private school in Berkeley, California. In 1960, while he was there, he met a Eurasian beauty named Nancy Hsu (徐乃錦); she was half-Chinese and half-German. Her grandfather was a hero of the Chinese revolution. In 1907, Hsu Hsi-lin (徐錫麟) had assassinated a high official of the Qing dynasty; for this, he was punished by having his heart cut out. Alan and Nancy planned to marry.

When they heard the news, Alan's parents were delighted; they believed their son had found a good match and hoped that marriage and a family would stabilise him. Nancy came from a good and educated family. The couple married at a Catholic church in California and went to study for a year at a school in Washington, DC. In 1961, Nancy presented Faina with her first grandchild, Chiang You-mei (蔣友梅), a daughter who was blond and had blue eyes. While the couple continued their studies in the U.S., they entrusted their daughter to Faina and CCK in Taipei; the two grew very attached to her. At 45, Faina was a young grandmother.

20th birthday of eldest son Chiang Hsiao-wen (蔣孝文) (Courtesy of "Academia Historica")

After six years in the United States, the young couple returned to Taiwan. In Taipei, they lived with Faina and CCK in the home in Changan Dong Road; the two couples went out together in the local area and ate snacks, like ordinary people. CCK demanded that Nancy, like his other daughters-in-law to come, did not take professional jobs; he wanted them to stay at home and be wives and mothers. They could do voluntary work. He wanted them to avoid the public eye and the media and "stay out of trouble". Nancy was well-educated and a polyglot; for her, this was a sacrifice but one she was willing to make. In Taipei, she threw herself into working with young people. (Zhou Yu-kou book, page 210).

During a visit to the Veterans General Hospital (VGH,臺北榮民總醫院), doctors informed Alan that, like his father, he suffered from diabetes. But he could not hold liquor as well as CCK. Thanks to the introduction of his father, he was given the post of an apprentice at Taiwan Power Company (臺灣電力公司); after a year, he was promoted to manager of a small office in Taipei. But, despite the promotion and being a husband and father of a baby girl, he continued to drink heavily. Sometimes, after such bouts, he verbally abused his wife. Once, when he was driving a jeep from Taoyuan (桃園) to Taipei with a friend, he drank while he was driving. His friend warned him of the dangers, to which he replied: "At 39, my grandfather was leading the whole country in the war with Japan. In his 30s, my father was famous all over China for his work in Gannan. But what am I in my thirties – a little official in Taiwan Power? The pressure on me is too great." (Faina biography, Wang Mei-yu, page 68).

On the evening of October 15, 1970, he attended a banquet and drank heavily; because of his diabetes, his blood sugar fell very sharply and he was very short of oxygen. He fell into a deep sleep at home which lasted long into the next morning. He carried sweets in his pockets; whenever he felt uncomfortable and feared his blood sugar was dropping, he ate one to restore the level. But, because he was asleep, he did not take one. When he did not wake up, his wife and family became alarmed and took him to a local hospital. The doctor saw how serious his condition was and had him transferred to the better equipped VGH. He remained in a coma; his wife stayed at his side night and day. Faina and CCK went to the hospital each day to see him. CCK whispered in his ear: "Baba is here, Baba is here. When you hear me, pinch my hand." (Zhou Yu-kou book, page 263)

He was in a coma for nearly a month. When he awoke, the doctor told the family that their son had lost part of his memory and would not be able to return to his condition before the coma. He had suffered damage to his brain. He remained in the hospital for almost five years. He lived mostly in a small house in the grounds, with his wife and daughter. According to his younger brother Eddie, his parents visited

Faina and granddaughter Chiang You-mei (蔣友梅) (Courtesy of "Academia Historica")

their eldest son each day during the coma. "Father held the hand of elder brother and whispered in his ear. 'Alan, Father has come to see you. Wake up!' Sometimes he would say 'Alan, Father says sorry to you.' This was because he felt responsible for passing onto him the diabetes he had inherited from the family of his (CCK's) mother. Every day Mother went and held his hand, in the hope that he would wake up. When she realised that the situation was not going to change, she simply could not accept it. This blow broke the heart of my parents (錐心之痛)." (Faina biography, Wang Mei-yu, page 69).

He was able to walk a short distance, with the help of his wife, a nurse or an aide. He was able to talk but not able to recover his full memory. In 1975, he was moved with his family to a residence on Yangmingshan (陽明山). With Alan incapacitated, Faina spent much time looking after his daughter You-mei; she became close to her and Grandfather.

Beloved daughter, unwanted marriage

Their only daughter Hsiao-Chang (孝章), whose English name was Amy, was born in Nanchang, Jiangxi in 1938, beautiful like her mother. Her character was different to that of her two noisy and troublesome brothers. She was soft and good-natured. Her parents adored her.

Bodyguard Weng Yuan said that Amy was very well-mannered to the security guards and wanted to be treated like an ordinary person.

At middle school, she refused the offer of a driver and escort and insisted on making the long journey from home by bicycle. CCK insisted that she eat a lunch box prepared at home. She stood at the school gate and took it discreetly; she did not want her classmates to see her as more privileged than they. (Weng Yuan autobiography, page 243)

She was strong-willed and had her own opinions. When she was 18, her father decided to send her to the U.S. to study. Nervous of sending a single girl to live alone in a foreign country, he looked for someone to look after her while she was there. He asked the "Defence Minister", Yu Ta-wei (俞大維) if his son Yu Yang-ho (俞揚和) would take the responsibility. The son, 40, was living in California and worked in business; he had been divorced three times. He was the son of Yu and a German lady whom Yu senior had met when studying in Germany; Yu Yang-ho was more than a dozen years older than Amy.

Amy went to study at Mills College, a private liberal arts college for ladies in Oakland, California, that had been founded in 1852. Neither Faina nor her husband imagined the outcome – their daughter fell in love with Yu and proposed to marry him. Her parents were

Top: Faina and daughter Hsiao-chang (孝章) in school uniform (Courtesy of "Academia Historica")
Bottom: CCK and daughter in military uniform (Courtesy of "Academia Historica")

bitterly opposed. "When the news reached Taiwan, there was an enormous shock (震撼) in the CCK official residence," wrote Weng (Autobiography, page 244). "When Amy insisted that she would go ahead with the marriage, CCK wept copiously (氣得淚眼直流) and could not speak for half a day."

The parents pointed to the big gap in age and Yu's record of failed marriages; they feared that the marriage would not turn out well. In addition, Yu was in business – CCK had always opposed his family going into business, fearful that they would use their political connections to enrich themselves. This has been a constant feature of Chinese political life, and is one reason why the public regard their leaders with disdain. "The exchange between power and money (權錢交易)" is the phrase common people use to describe this.

Yu Yang-ho was born in Germany in December 1924, when his father was studying there. He fell in love with a German piano teacher and Yang-ho was born. But her parents refused to give permission to marry the young Chinese. So Yu entrusted the baby, at 14 months old, to a cousin who took him back to China; he was raised by a relative there. He went to study in Hong Kong and passed an exam for the Air Force. In 1941, he was sent to the U.S. for training. In 1944, he returned to China to take part in the war with Japan. He took part in many operations; on the final one, he was shot down by a Japanese fighter and was injured after he parachuted to land. After the Chinese civil war, he emigrated to the United States; his wife refused to go and the two divorced. He later married two more women and divorced both.

Initially, Faina was strongly opposed Hsiao-chang to the proposed marriage. But she knew how strong-willed her daughter was; she feared that continued opposition would drive a wedge between her and her parents. "Faina was stuck between her husband and her daughter; it was very difficult for her. As a mother, she was unsatisfied with her daughter's choice. But good fortune and love were in the hands of her daughter. With the endless weeping and sadness of her daughter, she accepted her choice in this matter. She no longer opposed it; instead, she did her best to persuade her husband to accept it." (Zhou Yu-kou book, page 205)

In 1960, Amy returned to Taiwan for a visit; Father and daughter went to Sun Moon Lake (日月潭) for a weekend, which they spent boating and talking. She agreed not to marry until she had finished her studies. However, after she went back to the U.S., she and Yu went to Reno, Nevada and married there in secret. Neither Faina nor CCK were present at the ceremony. When he heard the news, CCK was eating lunch; he was so enraged that he pushed over the dining table. Faina wept; she was trapped between the anger of her husband and the

CKS, SML, Faina, Chiang Hsiao-chang and husband Yu Yang-he (俞揚和)
(Courtesy of "Academia Historica")

obstinacy of her daughter.

The couple settled in the United States. Their only child, a son named Theodore Yu Tsu-sheng (俞祖聲), was born in 1961. This dispute strained relations between Yu and his father-in-law; he rarely returned to Taiwan. But Amy went frequently to see her parents and brothers. She kept a low profile and out of the public eye. With the passage of time, CCK came to accept his daughter's choice. In 1970, during a visit to the U.S. as Vice-Premier, he made his first stop in San Francisco; he visited her and her husband. *TIME* magazine published a photograph of them having a meal together.

"I do not think of being a leader"

Faina's second son was Hsiao-wu (孝武), whose English name was Alex. His personality was similar to that of his elder brother; he also brought much grief to his parents, although he was more disciplined in his studies.

The two brothers often went out together to party and carouse; Alex also liked to carry a gun. On one occasion, his father even locked him in the house and forbad him from going out; only after pleading by Faina did he relent and open the door. His parents decided that, like

CKS (蔣介石), CCK, Faina and children (Courtesy of "Academia Historica")

his brother, he would do better to study abroad in a country where he would be treated like an ordinary person; they selected the Institute of Politics in Munich, Germany; he chose politics.

While he was studying there, he met Wang Chang-shih (汪長詩), the daughter of a Taiwan diplomat; the two fell in love. Alex's parents were delighted at the news, hoping that the marriage would make his life more stable. In 1969, they held the marriage ceremony at Amy's home in California. Since CCK was too busy with his work, Faina flew there to take part. It was the first time she had left Taiwan since 1949 and the first time she had visited the United States.

His wife gave Alex a daughter and a son, You-lan (友蘭) and You-song (友松). Unfortunately, the marriage was not a happy one; he had affairs and the couple argued often, sometimes in the family home. Finally, they divorced and Wang returned to Geneva, where her father lived in retirement; she was required to leave their two children in Taiwan. She later remarried.

After returning to Taiwan, Alex obtained a law degree from the Graduate School of Sino-American Relations of the Chinese Culture University (中國文化大學). From 1980 to 1986, he was president of the state-run Broadcasting Corporation of China (中國廣播公司). He had a busy social life, which included film stars and celebrities

Faina with son and grandchildren in home in Dazhi (大直) (Courtesy of "Academia Historica")

and members of the country's powerful intelligence community. In the spring of 1986, his father sent him to Singapore to work as deputy trade representative in Taiwan's mission there.

One reason was to cut his connections to the intelligence community. On October 15, 1984, two members of the Bamboo Union criminal gang shot and killed Henry Liu (劉宜良), a Taiwan dissident writer, in the garage of his home in Daly City, California. Those in the gang said that they had been ordered to carry out the assassination by Taiwan military intelligence. The media in the U.S., Hong Kong and even Taiwan said that Alex had been involved. He himself strongly denied the stories and said he had not been involved at all. In any event, going to Singapore was a good way to remove him from these suspicions.

Faina was sad to see him leave Taiwan; but, knowing the scandal surrounding the assassination, she could not oppose it and had to accept her husband's decision. She did not understand the complexities of the assassination, nor did she ask about it.

In Singapore, Alex married Michelle Tsai Hui-mei (蔡惠媚), the English teacher of his children. The daughter of a prominent Taiwan shipping family, she had studied at the Taipei American School; in 1977, at the age of 18, she started to teach English to Alex's two children. Their romance developed slowly. There was a gap of 14 years in age between them; he was a divorcee.

"It took Alex 10 years to win over Miss Tsai and her family," said Weng Yuan, the bodyguard (Weng Yuan autobiography, page 250). "Hers was a wealthy family from Taichung and did not need to mix with the Chiang family. Taiwanese basically had ambiguous feelings about the Chiangs. Add to that what they had heard about Alex's behaviour. So the family did not look well on the marriage. Without his dogged pursuit of the marriage, it would not have happened."

To help, CCK and Faina received the parents of Miss Tsai to their official residence for tea. To keep out of the public eye, the couple decided to marry in Singapore in 1986, rather than in Taiwan. She was the first native Taiwanese to marry into the Chiang family. The couple had no children of their own; she threw herself into taking care of Alex's two children by his first wife. The marriage stabilised his fiery character. "He worked very hard. Under the influence of his wife, he became a devout Buddhist. Initially, Faina brooded about his going to Singapore. But this change in her son calmed the worries in her heart." (Faina biography, Wang Mei-yu, page 81)

In January 1990, Chiang was posted to Tokyo as his country's top representative. It was not an easy assignment. Of all the peoples of

Asia, none love Taiwan more than the Japanese; it has strong support among the political class and the public. But Tokyo has diplomatic relations with Beijing, not Taipei; so government officials above the level of deputy section chief were not allowed to meet Chiang or his staff officially. The members of Beijing's embassy watched vigilantly to see if Japanese officials broke this rule.

In August that year, I interviewed him in a spacious meeting room in the Association of East Asian Relations, the name of the "embassy". He was surrounded by portraits of his grandfather and father staring down at him from the walls. They were powerful symbols of the weight of expectation on his shoulders. "I cannot compare myself with my glorious grandfather and father. I saw much of my grandfather during my youth. He never changed his daily schedule whatever happened, with a walk of at least 1,000 paces after dinner, meditation and prayer. Whenever I have a problem, I think of what my father and grandfather said in the past." Asked if he aimed for high office, he replied: "I do not think of it. My ideal is to serve Taiwan and my compatriots, not to take part in political demonstrations. I do not think today of being a leader nor will I tomorrow."

Disciplined student, successful businessman

Their third son, Chiang Hsiao-yong (蔣孝勇), known as Eddie, was

Chiang Hsiao-yong in 1964 (Courtesy of "Academia Historica")

born in Shanghai in 1948. Of the three, he was the easiest to raise. He was a good student; he followed Alan's footsteps into the Officers Military Academy (陸軍官校). Unlike them, he was a disciplined student who followed the strict rules of the academy and achieved good marks.

Unfortunately, during a training exercise, he injured his leg; this required months of treatment. He had to leave the Academy and transferred to Taiwan University (臺 灣 大 學), where he studied politics. While he was studying at the Academy, he met and fell in love with Elizabeth Fang Zhi-yi (方 智 怡), daughter of the director of the Expressway Engineering Bureau of the Department of Transport.

During Eddie's convalescence in hospital, Fang met members of the Chiang family. They liked her and accepted her as a daughter-in-law. In July 1973, during his third year at Taiwan University, the couple married. It was the only marriage of Faina's four children to be held in Taiwan. The couple invited members of the media to attend and also provided photographs of the event to others.

Eddie said that this had a political purpose. "At that time, the health of my grandfather was not so good and there were many rumours in political circles. So Grandmother and Father discussed this and decided to have a certain level of publicity of Grandfather at the

Top: Faina and CCK attend wedding of Chiang Hsiao-yong and Fang Zhi-yi (方智怡). (Courtesy of "Academia Historica")
Bottom: Fang Zhi-yi, Faina and Chiang Hsiao-yong (Courtesy of Fang Zhi-yi)

ceremony. It was a very happy event. We provided a photographer of the entire Chiang family, including CKS, to the foreign and domestic media. I was very honoured that my grandparents and parents were able to take part in and celebrate the wedding," he said. (Faina biography, Wang Mei-yu, page 89).

Fang Zhi-yi had studied computers and worked at a leading computer company. CCK did not want his daughters-in-law to work outside the home; so she resigned her job and became a housewife. She and her husband had three sons. Later, with the permission of her father-in-law, she set up a kindergarten.

Faina also worked in this field; she was chairwoman of the San Jun kindergarten (三軍托兒所), a private nursery run by a charitable foundation. She became increasingly shy and unwilling to appear in public. So she only attended events at festivals and to give awards.

After graduating from university, Eddie joined a KMT-owned company, the Chunghsing Electronic and Machinery Company (中興電工) and had a successful business career; he was chairman of several companies.

Main sources for this chapter:

Interview with Chiang Hsiao-wu by author in Tokyo 5/8/1990.

The Generalissimo's Son, by Jay Taylor (Harvard University Press 2000).

Biography of Chiang Fang-liang, by Wang Mei-yu (China Times Publishing Company of Taiwan, 1997).

My Years at the Side of Chiang Kai-shek and his Son, by Weng Yuan (China Times Publishing Company of Taiwan, 2015).

Chiang Fang-liang and Chiang Ching-kuo, by Zhou Yu-kou (Rye Field Publishing Company, 1993).

CHAPTER 6

She was Always Waiting for Him at the Door

In 1967, the family left the house where they had lived for 18 years. It was demolished because Changan Dong Road was being expanded. They went temporarily to a government guesthouse in Yangmingshan (陽明山) before moving to their new home in February 1968. It was Seven Seas (七海官邸), an official residence inside the Ministry of the Navy.

Seven Seas was a western-style villa constructed in the 1950s. The government built it as a guest house for the US military to accommodate visitors. But, in the event, it was little used and was empty most of the time.

It was practical, not luxurious; this suited CCK and Faina, who preferred simplicity to the trappings of power. CKS was delighted at the move. The new house was close to the residence in Shilin (士林) where he lived with his wife. It was inside the boundary of a military area protected 24 hours a day by armed soldiers; so, it was more secure than the former home, and easier for him to reach. He had long considered Changan Dong Road unsuitable and unsafe. Seven Seas had its own garden, where members of the family, friends and visitors could walk. It was set back from the road, so it was quiet.

Entrance of Navy Headquarters in Dazhi, Taipei (臺北，大直). Faina's second home in Taipei was inside this compound. (Photo by Mark O'Neill)

Faina would spend the rest of her life in this house. For her, the downside was that it was cut off from the world and inaccessible to the public. It was no longer possible to walk for a few minutes and reach shops, cafes and restaurants; from her old home, she could go shopping with her children, eat snacks and mix with ordinary people.

The new house was part of a sprawling complex of buildings in Dazhi (大直), on the northern side of the Keelung River in north Taipei that housed military facilities. CCK was "Minister of Defence" from January 1965 to June 1969.

In December 2014, the "Ministry of National Defence" (「國防部」) moved to a new eight-storey building in Dazhi. It houses the Air Force Command Headquarters, Navy Command Headquarters and Heng Shan Military Command Centre. It was built at a cost of NT$15.8 billion (US$516 million, using the 2019 conversion rate) and covers over 19.5 hectares; it includes office buildings and other facilities, such as a post office, barber shop, sports centre, conference hall and dormitories to accommodate the 3,000 military personnel stationed there.

The new house isolated Faina from the outside world. Access to the house was restricted to members of the Chiang family, trusted friends and people coming on official business. No-one could drop in casually.

As the health of CKS declined, his son took on greater responsibilities in the government, the party and the military. He became increasingly busy; he did not speak to his wife of what he was doing, nor did she ask. Much of it was top secret. During the late 1960s and early 1970s, Beijing and Washington were engaged in the secret diplomacy that led to the historic visit of President Richard Nixon to China in February 1972.

Recognising Beijing meant breaking relations with the "Republic of China" (Taiwan's official name); for CCK and his government, it would be one of their most dangerous moments since 1949. Would the U.S. abandon the Republic of China (RoC) completely? Faina learnt little of this. Like most people in Taiwan, she only knew what she read in the newspapers and saw on television. Her world was her husband, her home, her children and grandchildren.

The two worlds collided dramatically in the spring of 1970. Then "Vice-Premier", CCK visited the U.S. First, he went to San Francisco, where he had lunch with daughter Amy and her husband; this was an important family event, given the earlier bad feeling between CCK and his son-in-law.

Next stop was Washington where he received red-carpet treatment; he had long meetings with President Nixon, Secretary of State Henry

Kissinger and other top officials. On April 24, CCK was on his way to address the Far East-America Council of Commerce and Industry at the Plaza Hotel in New York. Just after midday, CCK got out of his limousine and walked up the stairs of the hotel. Two men with guns drawn jumped out from behind marble pillars at the entrance. As one opened fire, a detective pushed the gun away; the bullet flew past CCK's head and went through the glass of the door. Unperturbed, CCK entered the lobby and went to an upper floor and delivered his speech. Despite a police recommendation to leave the city, he insisted on completing his programme for the day.

On returning to his hotel, he called Faina and told her not to worry about him. It was in the middle of the night that Faina heard the news of the assassination attempt; it was a terrible shock and she could not sleep. When her husband returned from the U.S., she took her granddaughter Chiang You-mei (蔣友梅) to the airport to meet him. For her, it was a rare public appearance.

The would-be assassin was Peter Huang Wen-hsiung (黄文雄), an activist for Taiwan independence. He was a graduate student in journalism and sociology at Cornell University; like all men in the "RoC", he had served two years in the military. Many native Taiwanese opposed the martial law and one-party rule which the Nationalist government imposed on the island in 1949. Many demanded a democratic vote to choose their government. Huang served a five-year prison term in the U.S. – but did not return to Taiwan until 1996, after the statute of limitations for the assassination attempt had run out.

Faina was happiest as a mother and grandmother. She enjoyed the new home most when it was full of her children and their families. Her third son, Eddie, lived at home while he completed his studies at Taiwan University. He continued to live there for several years after his marriage to Elizabeth Fang (蔣方智怡) in 1973. The two had dated for eight years before they married.

Loving wife and mother

Elizabeth Chiang remembers Faina as a loving wife and mother. "I was not nervous before I entered the Chiang family. CCK did not demand that I give up work after marriage. I had worked as a computer programmer for IBM and then for CDC. But, after the marriage, it was I who felt the pressure. My boss felt the pressure. So I gave up work myself. Then I became pregnant. After our children grew up, I had to decide what to do. I opened a private kindergarten and ran it for 25 years. After my husband died, I continued to run it. Then, seven years ago, I left and started this charity to help single parents (Bystreams, 溪水旁關懷單親家庭協會)." (Interview with the author 19/12/2018)

CCK and Faina on hiking trips outside Taipei (Courtesy of "Academia Historica")

"After I entered the Chiang family in 1973, I saw how much Faina, my mother-in-law, loved and respected her husband and how much she loved her children. For her, the family was most important. She put no pressure on me. I felt the deep love between her and her husband. When he came home, she was always waiting for him at the door. He would call her 'Fang' (方) and she called him 'Guo' (國).

"When her husband was at home, she was always at home. They went together to parties and events, which was not common among leaders. She spoke to him in Chinese, with her Ningbo accent (寧波官話). I later saw in the archives Christmas and anniversary cards which she had written in Russian. She did not speak Russian to her children. When she met Russians who were performing ballet, she would use it. She studied Chinese hard. We were accustomed to her accent and could understand it," she said.

Faina was active and fond of sports. In Taiwan, she went with her husband for long walks and hiking in the countryside. Elizabeth Chiang said: "Faina liked golf and would play two-three times a week, leaving at 5:30am. She played at the Tamsui course. She played very well. Then she said that, one day, the people on the course ahead of her asked her to play first. She did not like to be made a fuss of, and decided to stop. My husband (Eddie) asked her to continue playing, but she declined. So she stopped playing.

"She swam when she was living in Xikou but not in Taiwan. She had many friends and was chairman of the San Jun (三軍) nursery. She took part in many activities with the children. Before, she used to play mah-jong. But I did not see her play. CCK asked civil servants to stop playing: so she stopped (even though she was not a civil servant).

"She had many friends. Often two or three wives would go to her house at 3:30pm and have tea with her. Most of her friends were Chinese. Sometimes, on the weekends, her husband went to Jinmen (金門) or Matsu (馬祖). She did not go with him. She went with friends to the United Hotel (統一大酒店), to have dinner and listen to music. She did not join her husband in his public appearances. She had no interest in politics. The house in Changan Dong Lu was too old, a traditional Japanese house. She had good relations with Madame Soong Mei-ling. They went together to Christmas parties and school activities. They took part in many meetings of women's associations.

"In the early years, she used to cook. Later they had a cook, Sister Ah-Bao (阿寶姊), who had joined the family in the mainland when she was 17. She came with the family to Taiwan. She was like a member of the family. The family always ate Chinese food. Faina's favourites were borscht（羅宋湯）and cabbage and meat. Her car was a Buick.

"She read many books, in English. She read English newspapers. I did not see her read Russian books. They were very hard to get in Taiwan at that time. She was not so interested in television series; she put them on to have a sound in the house.

"She was a very good mother-in-law. She did not criticise me. She did not get angry. She did not tell me how to bring up my children. We would tell her about them. She let her own children develop as they wished. I remember in 1996, when her own health was declining. She told me to take care of my husband and not to worry about her. I was impressed. I thank her for the way she treated me as a daughter-in-law," Elizabeth said.

Son Eddie said that one reason his mother stopped playing golf was her asthma. (*Biography of Chiang Fang-liang*, Wang Mei-yu, page 51). For the same reason, she had to stop smoking, a habit she had enjoyed for decades, on the advice of her doctors, he said.

"Initially, in the house in Dazhi, the atmosphere was warm and busy. Later, the children grew up and formed their own families. Some went to live abroad, so the house became quieter," he said. She helped to bring up her first grandchild, Chiang You-mei. Both Faina and CCK had a particular affection for You-mei; she was only nine when her father fell into a coma and so she spent more time with her grandparents. For her senior high school, she left Taiwan to study in the U.S., which CCK found hard to accept at first. "While she was studying abroad, You-mei always found time to write to the grandpa who doted on her." (Weng Yuan autobiography, page 287). "Each time the letter arrived, Faina was extremely happy and took the letter for her husband to read." Later You-mei went to the U.K. to study; she continued to write letters. During one winter holiday, she came back to Taiwan and visited her grandparents. At that time, CCK was in poor health; he was confined to bed for much of the day. Delighted, he asked You-mei not to return to U.K. but to stay in Taiwan and stay with Grandfather. "'Impossible', she said. 'I have to go to Britain and continue my research.'" (Weng Yuan, autobiography, page 287)

Spartan and thrifty

In his autobiography, bodyguard Weng Yuan gives a vivid description of life in the new home, where he went to work in 1978. He contrasted it sharply with the residence in Shilin of CKS and SML. The land area of Seven Seas was just a tenth of that of Shilin. Leafy and quiet, it had been built more as a villa, for someone to stay in for a short time, than as a family home. It did not have the facilities of Shilin, such as a car repair shop or an area for the domestic staff.

Faina and CCK insisted on thrift and simplicity, with a daily budget for food of below NT$1,000 (US$33, using the 2019 conversion rate), a tenth of that in Shilin. The staff used the equipment and electrical appliances in the kitchen they found on their arrival and did not replace them. The cook and housekeeper who implemented these Spartan regulations was Sister Ah Bao (阿寶姊). She was born in Dinghai (定海縣) in Zhejiang, the native place of the Chiangs; she started working for the family in the mainland. She and Faina got on very well; she worked for them in the Changan Dong Road home and stayed with the family her whole life. Faina and Ah Bao had similar ideas about frugality. When there was old and fresh fruit in the refrigerator, Ah Bao would always serve the old pieces first; if someone asked for the fresh fruit, she said that they must eat the old fruit first and not waste money. She resisted attempts by family members to replace the old equipment, saying that it could still be used. For more than 20 years after the family moved in, they bought nothing new for the kitchen. Ah Bao worked in the house 365 days a year and took no holidays. She looked after everything bought for the house, including food, clothes and household items.

Most remarkable for the First Family were the rules for food; this was a family whose orders could not be questioned. Food is one of the most important elements of Chinese life and society. Spending several thousand NT dollars a day on food would have been easy; most people would have considered it acceptable for someone as senior in the government as CCK and who needed to entertain guests. At Shilin, the monthly catering bill was NT$200,000 – 300,000 (US$6,500-US$9,800, using the 2019 conversion rate); this allowed for banquets and preparation of expensive and specialty foods, for guests or the residents. But, in Seven Seas, there were rules for what would be served to visitors, in three grades; Faina did not order more than what she thought the guests would eat. If the cook spent more than NT$1,000 (US$33, using the 2019 conversion rate) on buying food, he or she had to show her the bill to check.

The biggest spreads were for birthdays and anniversaries in the family, especially in 1984 to mark the couple's 50th wedding anniversary. "CCK and his wife paid particular attention to this and invited family and friends to the event." (Weng Yuan autobiography, page 284). "That day there were two tables. That kind of banquet was very rare in Seven Seas. We could see that CCK wanted to show the importance he attached to the anniversary. He wanted to show his thanks to his wife for a life of pain and hard work. I saw with my own eyes CCK's softness toward Faina. It moved us." Weng also described what he saw several times while he was on duty outside CCK's bedroom at night. "He entered the bedroom of his wife and went to her side. They held the two hands of the other and looked at each other for a long time. Then, without saying a word, CCK left and returned to his bedroom." (Weng Yuan autobiography, page 283) CCK did not spend time on

Faina celebrates her birthday with three children. (Courtesy of "Academia Historica")

his wardrobe. "In his clothes, CCK was never very particular. He had several pairs of suits and ties which he wore all the time." (Weng Yuan, autobiography page 280) "He and his father both liked to wear old clothes. Where he was different from his father was that he did not care about the brand. As long as the clothes were practical, that was fine."

Many of his clothes came from daughter Amy who brought outfits from California on her visits home; she brought fabric from which her mother had clothes made. Her departures from Taipei were emotional for her parents, who did not want to see her go.

Faina was similarly frugal in her habits. "She was influenced by CCK and was thrifty and economical. She was a traditional person who saved things at home." (Wang Mei-yu, Biography of Chiang Fang-liang, page 95) This frugality extended to her wardrobe. She preferred to wear the same sets of clothes and rarely bought new ones. Since she rarely made public appearances, she was not under the pressure of impressing audiences or the media. To have her hair done, she paid regular visits to a salon outside.

Death of the Generalissimo

Just before midnight on April 5, 1975, the heart of Chiang Kai-shek stopped beating; he was 87. A heavy storm with thunder and lightning struck the whole of Taiwan, from north to south; everyone saw it as a sign that something momentous had taken place. Cinemas and places of entertainment closed for a month. More than two million people filed past CKS's casket during the five days it lay in state at the Sun Yat-sen Memorial Hall in central Taipei. The final service for him was held there on April 15, before his body was driven to a modest sarcophagus in the mountain retreat of Tzuhu (慈湖), 55 kilometres southwest of Taipei. Hundreds of thousands of people lined the route.

Faina knew this was a very important moment in her life and that her husband would in time take over from his father. While "Vice-President" Yen Chia-kan (嚴家淦)succeeded CKS as "President", the standing committee of the Kuomintang unanimously chose CCK as the chairman of its central committee. So he would become "president" when Yen completed his term in May 1978. For Faina, this meant her husband's workload would only grow heavier and she would see even less of him.

Soong Mei-ling, the widow of CKS, remained active in public affairs after the death of her husband. In 1975, she emigrated to the U.S.,

25th wedding anniversary of CCK and Faina (Courtesy of "Academia Historica")

where the family had a 14.6-hectare estate in Nassau County, New York. She rarely returned to Taiwan. But everyone, including the media, called her "Madame Chiang" (蔣夫人) and continued to until her death in 2003, aged 105, in New York.

Faina never held that title, even when her husband became "president"; it was reserved for Mother-in-law. The contrast between her and Faina could not have been sharper. SML loved public life and held strong opinions about politics; Faina avoided the spotlight and did not speak about politics. On May 20, 1978, CCK duly became the "President"; he was 68.

After taking office, CCK decided to stay in their Seven Seas home. Within a military compound, it was certainly secure enough as the residence of the "president". But it was modest for a head of state; most would have chosen somewhere more palatial and luxurious – or built a new one. Faina also felt at home there. She and her husband both preferred its modest surroundings to the trappings of power. For his office, CCK used the same suite as his father.

Diabetes

In the 1960s, CCK was diagnosed with diabetes, which he had

inherited from his mother. He in turn passed it on to his eldest son, Alan. In the later years of CKS, CCK was driven each morning to Shilin to check the state of his father's health. A nurse gave CCK a blood test, to see the level of his blood sugar; according to this level, the nurse gave him an injection of insulin. Then he had breakfast and went to his office.

After Alan fell into a coma, the family paid even more attention to CCK's diabetes. The Veterans General Hospital (VGH) sent two doctors to Seven Seas; they were on duty 24 hours a day and made regular tests of his blood sugar. But, in the 1970s, his condition worsened.

Weng Yuan said that the main reason was that CCK did not follow the dietary advice of his doctors. "He ate whatever kind of food he liked, especially when he was on inspection trips outside (Taipei). If he felt hungry and saw something nice on a street side stall, he would eat it, regardless of his blood sugar." (Weng Yuan, autobiography page 303). These trips were a feature of CCK's leadership and distinguished him from his father and most Chinese leaders. Their visits were tightly scripted, with heavy security; the people – and usually the dialogues that were to take place – were carefully vetted in advance. So it was, and is, hard for leaders to find out anything but the "official truth" or what the subordinates who arranged the visits wanted them to hear.

CCK and Faina (Courtesy of "Academia Historica")

CCK knew that the Nationalists did not have a democratic mandate from the Taiwan public; he wanted to compensate in part by spending weekends outside Taipei and listening to ordinary people. He did not announce the itinerary; usually wearing a baseball hat, he tried to have an honest conversation and learn the real opinions of those he met. In Taiwan, as everywhere, a good way to create a bond is to share food with others and enjoy the delicacies of each village and region. CCK also drank liquor with his hosts. All this made good media photographs and showed a leader close to the public. It was good for democracy – but the worst diet for his diabetes.

Faina knew well the bad effect of this on her husband's condition; but he would not take her good advice about it. She could only watch helplessly. The only person he would listen to was their long-time housekeeper Sister Ah Bao. "Her authority far exceeded that of Faina. Whenever she told him to eat or not to eat something, he knew that there was no room for negotiation or discussion." (Zhou Yu-kou book, page 267)

In 1981, to control his rising blood sugar index, the doctors increased the insulin injections from once to twice a day, before breakfast and before dinner. (Weng Yuan autobiography, page 303) But he continued to ignore the advice of the doctors; his strong temper made them unable to control him. When he demanded an ice cream – a "banned" food because of its high sugar content – the staff had no alternative but to provide one.

Working from Home

As CCK's health deteriorated, his home became his office. After becoming "president", he complained to his doctors of discomfort in his feet and legs. In January 1980, surgeons at the VGH operated to remove a cancer in his prostate. That year he reported constant pain in his leg and needed sedation to sleep. In 1982, he needed an operation at VGH for a retina problem. By the end of 1983, he could barely walk, but refused to use a wheelchair. He also suffered from severe headaches.

So it was that Seven Seas became his office; on most days, he spent only one hour at the "Presidential Palace" and held other meetings at Seven Seas. In February 1984, he had a hospital bed installed in his second-floor room there; it could be adjusted to the sitting position, as required. By the end of 1984, his eyesight had deteriorated to the point that he needed his staff to read to him.

He met people in his bedroom or, if the numbers were large, in a reception room. Like many leaders, CCK did not want the Taiwan public or the outside world to know how serious his condition was. Sometimes he had to stay the night at the VGH. He ordered his staff to put a fake CCK in his car and drive to Seven Seas and then for the fake CCK to return to the hospital the next morning. (Weng Yuan, autobiography page 263). He disliked having to stay in the VGH. When he was there, he often did not use his real name; instead, he wrote that of one of his bodyguards. One day Weng Yuan went there for a blood test himself and discovered that his name had been "borrowed" by CCK. So, he had to register as "Weng Yi-yuan" (翁一元). (Weng Yuan autobiography, page 264). In the name of security, the hospital staff went along with this practice.

Faina's health was also declining; but she was better off than her husband. She suffered from insomnia and shortness of breath; she and her husband slept in separate rooms. She saw the health of her husband deteriorating; she was constantly in his room to take care of him.

One of the most important meetings in Seven Seas came in the early morning of December 17, 1978. U.S. ambassador Leonard Unger arrived there about 4am to give CCK and his aide James Soong (宋楚瑜) an official statement from the U.S. that President Jimmy Carter would recognise the People's Republic of China a few hours later. Soon members of the cabinet and other senior officials arrived at the house for a meeting, over breakfast, to discuss this momentous announcement.

In the end, the "RoC" weathered the crisis well. Adroit lobbying in Washington and strong sympathy among the American public and political class for the "RoC" resulted in the Taiwan Relations Act, enacted by the U.S. Congress in April 1979. It permitted de facto diplomatic relations and said that more than 60 treaties and agreements between the "RoC" and the U.S. before 1979 were still valid. It said that the U.S. would continue to provide defensive arms to the "RoC" and resist coercion against the security or social or economic system on the island. In 1980, foreign investment increased over the precious year; the economy continued to grow. The normalisation of relations between Beijing and Washington did not lead to the isolation of the "RoC" from the U.S., Japan and other major countries. It had lost its seat at the United Nations and its major diplomatic allies; but it established strong business and non-official ties with them and its economy continued to grow strongly. The nightmare had been averted.

Between 1981 and 1984, a frequent visitor to Seven Seas was James Lilley (李潔明), head of the American Institute in Taiwan (AIT), his country's unofficial embassy. The two men got on very well. After spending the first 20 years of his life in China, Lilley spoke fluent

Mandarin. His father was an executive for Standard Oil and moved to China in 1916; he hired a Chinese nanny who talked to the young James in Mandarin.

Before going to Taiwan to head AIT, Lilley had spent nearly 30 years in the Central Intelligence Agency, with assignments in Laos, Japan, Hong Kong, Taiwan and mainland China. From April 20, 1989, he served as U.S. ambassador in Beijing, just in time to witness the student-led protests in Tiananmen Square. Wearing red check trousers, he cycled around the square; he got off and talked to the students, to understand what was going on. When foreign journalists spotted him, he asked them not to report his presence; the Chinese government would see it as interfering and accuse him of instigating the protests. Other foreign ambassadors did not dare to do such a thing.

In April 1985, the VGH surgeons gave CCK a pacemaker, which helped to stabilise his heartbeat. He began to use a wheelchair in public. His third son, Eddie, became one of his most important aides. Each Tuesday and Friday, he gave a report and analysis to his father on developments in the "RoC"; he read him newspapers and reports. He accompanied him to visit the trips on the front-line island of Kinmen (金門). Eddie became CCK's eyes and ears and spokesman. He took on many responsibilities in the house, including organising the security staff and what meals the staff should make for his parents. "Faina greatly relied on him." (Biography by Wang Mei-yu, page 92)

Succession

From the time he became "president", CCK was in physical discomfort that grew increasingly severe. In another country, he might have retired and given his place to someone in better health. He did not do so for two reasons. One was that he was the son of CKS, the man groomed by his father for years to succeed him; this was what thousands of their followers expected. The other reason was a strong sense of mission, to achieve certain objectives before his death. He knew that he did not have many months to live. This meant implementing major decisions that only he could take.

One was to pick his successor. In 1983, CCK decided that his successor would be a native Taiwanese, to reflect a new era of democratic politics. The mainlanders who came in 1949 and their descendants accounted for only about 15 per cent of the population. CCK decided that this group could not continue to dominate the government as they had during his era and that of his father. He ruled out his three sons succeeding him or occupying a senior position in the Nationalist party.

The man he chose was Lee Teng-hui (李登輝), a specialist in agriculture who had studied at the Taiwan University and Iowa State and Cornell Universities in the United States. Between December 1981 and May 1984, Lee served as chairman of the Taiwan Provincial Government (臺灣省政府). CCK chose him as his "Vice-President"; his term began on May 20, 1984.

Other major decisions were: the abolition of martial law the government had decreed in Taiwan, Penghu, Kinmen and Matsu in May 1949; reform of its political institutions to reflect the fact that they exercised control only over these four areas and not mainland China: and lifting a ban on new political parties and media not controlled by the government and Kuomintang. His objective was that the people of the "RoC" could choose their leaders through elections. These reforms faced fierce opposition from conservatives within the Kuomintang, the military and security services; they argued that a democratic "RoC" would be chaotic, hard to govern and could lead to a declaration of independence that would trigger a war with mainland China. This opposition included stepmother Soong Mei-ling. Aged 85, she returned from her gilded retirement in Long Island in October 1986 for an extended stay at her Shilin residence. From there, she campaigned against the reforms she saw CCK undertaking; she argued that they would lead to chaos and disorder.

Lee Teng-hui (李登輝) meets Faina in Tzuhu Mausoleum of CKS and CCK (慈湖陵寢). (Courtesy of "Academia Historica")

CCK knew that he alone had the authority to push through these reforms; he knew that, if he did not, his successor would probably be unable to. The conservatives were powerful and controlled important parts of the government; they would block changes that did not have the imprimatur of the Chiang family. That is why CCK enacted these reforms during the last 12 months of his life. He ordered the lifting of martial law on July 15, 1987, after 38 years; the same day a new Security Law went into effect. On January 1, 1988, on his orders, the government lifted the restrictions on the number of newspapers and the pages they could print.

He permitted the establishment of the Democratic Progressive Party (民進黨) in September 1986, even though it was at that time illegal under martial law. In 2000, its candidate won the presidential election and became the leader of the "RoC", the first since 1949 not a member of the KMT; this victory was a legacy of CCK.

In 1987, CCK spent most of his waking hours in bed or the wheelchair. On Christmas Eve that year, he and Faina attended a family dinner hosted by SML at her home in Shilin. Present were their sons Alan and Eddie with their families and CCK's adoptive brother, Chiang Wei-kuo (蔣緯國) and his family.

On January 6 1988, Faina suddenly fell ill; her doctors advised her to go to VGH. Weng Yuan said that this worsened an already tense situation within Seven Seas. "She was suffering from a weakening of her lungs brought on by asthma. It was an emergency. The ambulance was waiting at the entrance, ready to take her to VGH at any moment." (Weng Yuan, autobiography, page 340). She did not want to go. Eddie asked to, but in vain. Only when her husband offered to go with her did she agree; they stayed several days. Eddie said: "The reason why she did not want to go to hospital was that she did not want to leave her home and her husband. The older she became, the more she relied on him. Although both were in poor health and only exchanged a few words during a day, having him by her side was her anchor. Although the days were quiet, she felt very secure and comfortable." (Biography by Wang Mei-yu, page 97). Alex Chiang recalled that, during his father's late years when his health was deteriorating, he told him that the children must take very good care of their mother. "She has sacrificed everything for this family," said CCK. (Zhou Yu-kou book, page 230)

Passing

On January 13, 1988, Faina had returned from VGH and was resting in bed from her treatment. That morning CCK woke early, as was his habit. He felt uncomfortable and wanted to vomit. His doctor was summoned but could not relieve his discomfort. About 40 minutes after midday, he began to vomit large amounts of blood. The doctors and bodyguards were alarmed; they called for son Eddie to come.

Working at the house that morning was bodyguard Weng Yuan. He described how staff, doctors and nurses did their best to save their ailing CCK; all could see how critical the situation was. He said that they did not wish to tell Faina of her husband's condition, because they feared that she could not bear the shock. "Everyone was mobilised ... but no-one can fight with destiny" (Weng Yuan autobiography, page 347)

Despite the best efforts of everyone, CCK's eyes closed and his expression was serene. Doctors confirmed the time of death at 3:50pm; he was 77. Eddie called top officials of the government to come urgently. "When CCK was coughing a lot of blood, the residence was in great confusion. But no-one had dared to go into the neighbouring room and tell Faina. Her health was not good. She had only returned a few days earlier from the VGH and was wearing an oxygen mask. The ambience in the house was becoming more and more tense. After calling the government leaders, Eddie could not conceal the matter any longer and told Mother. This was a bolt from the blue. She could not accept it." (Biography by Wang Mei-yu, page 99)

She and CCK had been married for 53 years. The leaders did not go to see Faina but could hear her weeping in the next room.

Farewell

On the morning of January 30, Faina and her four children saw CCK for the last time. His casket had been lying in state at the National Revolutionary Martyrs' Shrine; 1.2 million people had filed past it to pay their respects. Then the five of them and a small number of friends, including Singapore Prime Minister Lee Kuan Yew (李光耀), accompanied the body to a small house in a town called Touliao (頭寮) , near the burial place of his father at Tzuhu (慈湖). Their final resting place – in Taiwan or the mainland – remained unresolved even in 2019, with no end in sight. A million people lined the route. At 9am, almost the entire population of the island stopped what they were doing, to pay their last respects.

Faina attends funeral of her husband. (Courtesy of "Academia Historica")

Main sources for this chapter:

Interview with Elizabeth Chiang (蔣方智怡) in Taipei 19/12/2018.

The Generalissimo's Son, by Jay Taylor (Harvard University Press 2000).

Biography of Chiang Fang-liang, by Wang Mei-yu (China Times Publishing Company of Taiwan, 1997).

My Years at the Side of Chiang Kai-shek and his Son, by Weng Yuan (China Times Publishing Company of Taiwan, 2015).

Chiang Fang-liang and Chiang Ching-kuo, by Zhou Yu-kou (Rye Field Publishing Company, 1993).

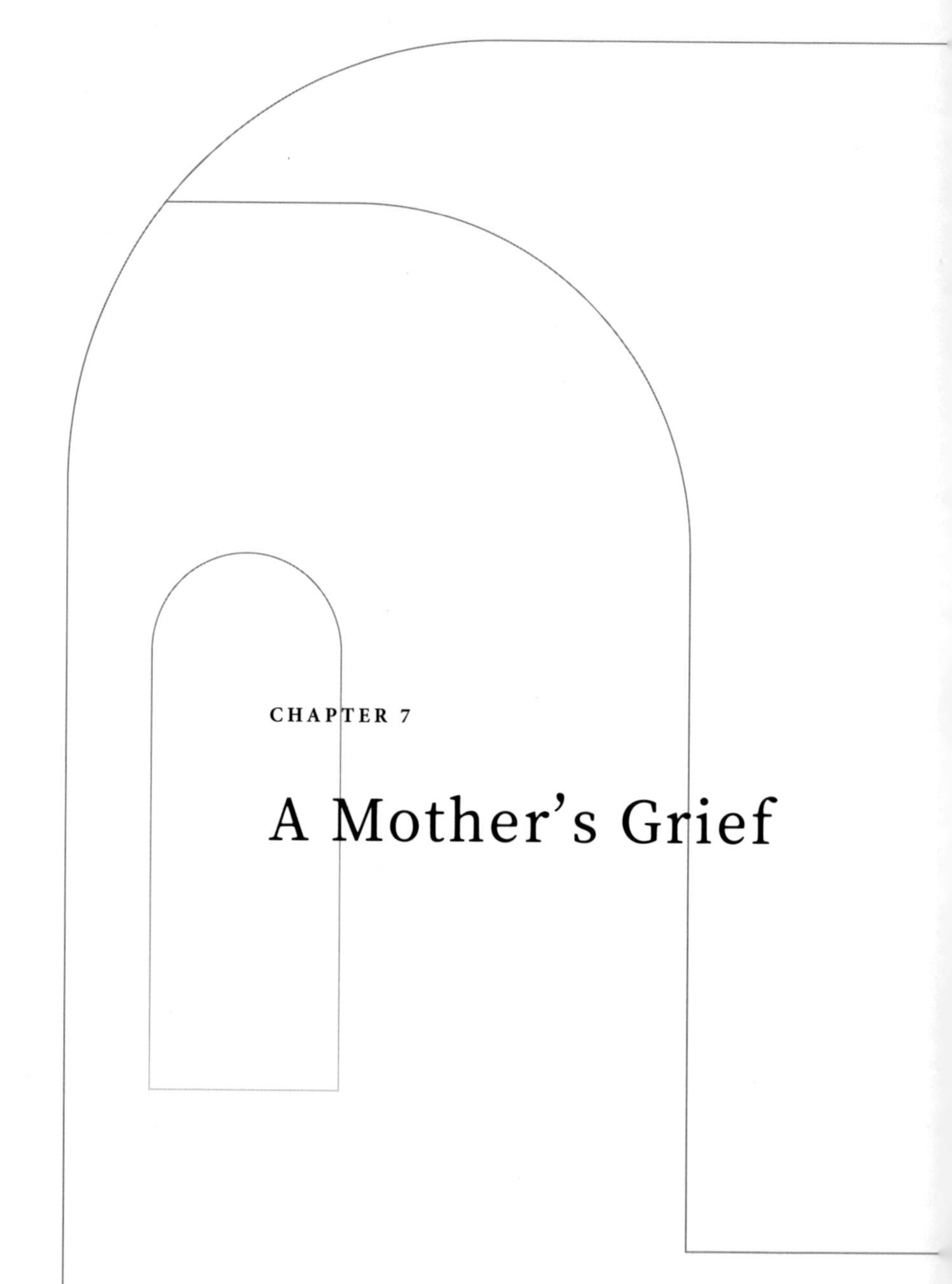

CHAPTER 7

A Mother's Grief

Faina's life changed dramatically after the loss of her husband. The man who had been her support and closest companion for 53 years had left her side. Outside of him, her social circle was limited – her children, grandchildren and other members of the Chiang family and old family friends.

For 10 years, Seven Seas had been the nerve centre of the government, from which CCK directed affairs. The house was full of senior officials and their staff coming to attend important meetings. From now on, they would not come, except for courtesy visits during Chinese New Year to see the widow of the leader. The action moved to the home of his successor.

The regular visitors now were Faina's children and grandchildren and close family friends. She was less and less inclined to leave the house. Fearful that she was becoming isolated, her family urged her to go out and mix with the world; she scarcely followed their advice. Her children invited her to move to California and settle there with them.

After the death of their father and his choice of Lee

CCK, Faina and their four children at home (Courtesy of "Academia Historica")

Teng-hui as successor, the three sons saw that they had no political future in Taiwan. The island was entering a new era of representative government. So they had to decide their future. The youngest, Eddie, chose to emigrate with his wife and three children to Montreal, in April 1988. He told a Taiwan magazine that he felt that this family and the modern history of China had been linked together, but that now was time for a pause. It was an opportunity for the family to start a new life as ordinary people, in a place where few knew who they were. After living for a period in Montreal, the family moved to San Francisco, where they were close to sister Amy who lived there. Eddie's departure left Faina with only one of her children in Taiwan, her eldest son Alan. Alex was in Singapore as director of the government's trade mission there.

Faina and her husband had few assets – hard to believe for the son of CKS who had occupied senior posts in the government for more than 20 years. CCK had had many opportunities to make money, legally and illegally. The couple possessed no property of their own. Seven Seas belonged to the government; Faina lived off her own and her husband's pensions. She also had money in a bank account which she had saved over the years.

"Since he entered government, CCK kept a distance with people in business and did not want contact with them. He was indifferent to money. Yu Kuo-hua (俞國華) who was with him for a long time said that, after his death, he had no personal assets. The properties he used were all public. He gave all his salary to Faina to manage. He had no concept of money. When he received allowances, he gave them to his bodyguards or donated them to poor people." (*Biography of Chiang Fang-liang*, by Wang Mei-yu, page 94).

So there was no question of Faina moving anywhere. In any case, Seven Seas was where she wished to stay. That was the centre of her life and her family; it held the memories of her husband. The government was happy to let her remain there, out of respect for her position as the widow of the leader.

During the last weeks of CCK's life, the doctors, nurses and staff at Seven Seas were nervous of Faina's condition and how she would deal with the loss. She suffered from asthma and a weak heart. But she showed a strength and resolve they had not expected. "Faina was strong in the midst of her deepest grief. She stood up with a straight back. Starting from January 14, 1988, she made a spiritual pact, to continue her life in the world of Chiang Ching-kuo. (*Chiang Fang-liang and Chiang Ching-kuo*, by Zhou Yu-kou, page 273)."Faina kept her husband's bed, the chair next to it and the small table in the bedroom. Each morning, after she got up, she opened the door and went into the room as before. There it was as if a smiling CCK was talking with her … In the tomb in Touliao, she could sit quietly for many hours on the marble next to his resting place and recall the good times she had had with her husband." (Zhou Yu-kou, biography, page 273)

Her daughter-in-law Elizabeth Chiang said that, when CCK died, Faina wept. "But, in front of other people, she did not let people see her feelings. Sometimes, her driver took her to Keelung when she would sit in front of a bronze statue of her husband (基隆八斗子碧砂漁港). She would talk to people there and then come back. She liked that. In her later years, she gave me a task. She wished to be buried next to her husband, whose grave is in Touliao (頭寮) … I told her that there was no space for her to be buried next to him: how about the next room? 'No, I must be next to him,' she said. So I had another suggestion; I did not fear that she would be angry. 'After you leave, you can be cremated and the ashes laid next to him.' This she accepted. So this is what we need. If you go there, you will scarcely see it. It is next to CCK's grave. There is nothing to say what it is." (Interview with author, Taipei 19/12/2018)

Faina was greatly supported by her staff in Seven Seas. Sister Ah Bao stayed on; she devoted her life to the family and the house. She said that her food and accommodation were taken care of and she did not need to worry about her livelihood. "I have stayed on in Seven Seas, out of a sense of gratitude (*My Years at the Side of Chiang Kai-shek and his Son*, by Weng Yuan, page 294).

Another person who looked after Faina was a nurse named Chen (陳); she had helped with the birth of You-mei, the daughter of Alan. Faina took a great liking to her and invited her to work with the family. She stayed for more than 30 years, including after the death of CCK, more as a companion than a nurse.

Christian faith

Elizabeth Chiang said that one important pillar that sustained Faina was her religious faith. "She was a Christian, like her husband and her in-laws. The family said grace before meals. She read the Bible. Her faith was a very big part of her life. It enabled her to deal with the deaths in her life. She believed that God had chosen the time of life for each person. She was sad but not angry. She did not ask why. She knew that God had arranged it. She said this to comfort me after my husband passed away. She quoted these verses from the Philippians, chapter three, verse 13/14: 'Brothers and sisters, I do not consider myself yet to have taken hold of it. But one thing I do: forgetting what is behind and straining toward what is ahead, I press on toward the goal to win the prize for which God has called me heavenward in Christ Jesus.'" (Interview with author, Taipei, 19/12/2018)

The family pastor was a Baptist minister Chou Lian-hua (周聯華牧師) who often came to the house. He ministered to the Chiang family for 40 years. Chou's family hailed from the Cixi (慈谿) district of Ningbo and he, like the Chiangs, could speak Ningbo dialect. He was born in Shanghai in March 1920 and educated at schools in the city. In February 1949 he left Shanghai for the United States, expecting to return. After receiving his doctorate in Theology at the Southern Baptist Theological Seminary in Kentucky he was invited to become a teacher in the U.S. but chose instead to go to Taiwan in 1954 and become a minister.

Invited to preach on Sundays at CKS's the Songs of Praise Church in Shilin (士林凱歌堂) Chou became its regular minister. This was a private church; all the attendees were there by invitation. Chou was discreet about his work there; he did not tell others he preached to the Chiangs and had no name card revealing so. (*Memoirs of Chou Lian-hua*, page 198).

Chou said that CKS was very devout and attended the service every Sunday, unless he was sick or had an urgent work commitment. He said that CCK had been baptised a Christian in Chongqing during

World War Two. "He read the Bible and often quoted passages from it in his speeches. He did not often come to Songs of Praise Church; but he did at Christmas and always brought all the family" (*Memoirs of Chou Lian-hua*, page 222).

The main reason for CCK's absence on Sundays was that he usually spent the weekends on inspection tours around the island. "After he became sick, he of course could not attend. Minister Chou said that, after Alan fell ill, he used to visit him; when he returned home, he prayed for him. "The children of great men cannot always be great men (*Memoirs of Chou Lian-hua*, page 229).

Double blow

Just 15 months after losing her husband, Faina suffered a second tragedy – the passing of her first-born son. On April 14, 1989, Alan died of cancer of the nose and the pharynx in the Veterans General Hospital. He was just 53. He was the child who had accompanied his parents throughout their long odyssey, from Sverdlovsk to cities all over the mainland and then Taiwan. He carried for his mother so many memories, both sweet and bitter. "She wept alone. Hers was the tragedy of a mother saying goodbye to her son (Wang Mei-yu, biography, page 102)

Number two son Alex was serving in the representative office in Tokyo. He invited her to move there and live with him and his family. She was touched by the invitation but declined it, not wanting to leave Seven Seas. "In Dazhi, each spot, each piece of wood, each corner and each brick carried the sweet memory of her life with her husband (Zhou Yu-kou, biography, page 327).

After the death of Alan, Alex applied to the government to be transferred back to Taiwan, so that he could take care of his mother. She was delighted at this news. Alex felt remorse over his past behaviour; relations with his father had been fractious and he had brought grief to his parents. Now was the moment to make up and return to his mother some of the love and attention she had lavished on him. He had been living abroad since spring 1986, first in Singapore and then Tokyo. The new "president" appointed him chairman of the state-owned China Broadcasting Corporation (中華電視公司董事長).

Faina was delighted; she had greatly missed him – although she understood the reasons why he had been sent abroad. He promised his mother that he would never leave her again. He returned in 1991. On June 30, the day before he was to start his new job, he was preparing a speech to the staff on his opening day; he went to the VGH for tests. Early the next morning, a telephone call interrupted the silence of

Seven Seas. It was a member of the hospital staff, with astonishing news. Alex had died during the night, at the age of 46. He had succumbed to a disease of the pancreas, which doctors had discovered during his examination; this was due in part to his having taken pain-killers over a long period.

That afternoon, Faina was taken in a wheelchair to VGH for final look at her son. She wept for a long time. This moved the staff of the hospital as well as people across Taiwan when it was shown on the television news that evening. They saw the elderly mother in a wheelchair, her head bowed and soaking the tears with a handkerchief.

In three and a half years, she had lost her husband and two of her sons. She summoned all her strength to deal with the grief. "She knew that weeping and grieving could not solve the problem. Even less did she want her grief to influence the feelings of her children. Using a great level of self-control, she did as she had during her youth – put out her chest, stood up tall and told her children and grandchildren that everyone should work hard with their family and professions. She would stay on her own in Dazhi, remembering the laughter and words of her husband, give thanks and have a peaceful and happy life (Zhou Yu-kou, biography, page 326)

Third son Eddie was living with his family in San Francisco. He returned regularly to Taiwan. While he was there, he visited his mother for lunch. Another regular companion was Nancy Hsu (徐乃錦), Alan's widow. Faina's daily routine was to read the English-language newspapers and watch news on television. She was well informed on current affairs but found the Chinese-language newspapers hard to read. She read books and looked through family photo albums but did not speak much.

Eddie said: "Perhaps because she had been silent for a long time, she was not so much interested in the visit of friends. When I raised the issue of people coming to see her, she always said: 'Thank you for your good intention.' But she did not want to see them." (Wang Mei-yu, biography, page 103)

Mission from home

On the evening of December 26, 1991, the flag of the Soviet Union was lowered from the mast over the Kremlin in the centre of Moscow for the last time. It was replaced by the tricolour used by Tsarist Russia before the revolution in 1917. The Soviet Union ceased to exist; in its place sprang up 15 independent republics, among them Belarus, Faina's mother country. The Cold War ended. For the first time, it became possible for Taiwan to have relations with these countries –

and for Faina to visit her native place and Sverdlovsk where she had met and married her husband.

The end of the Cold War also lifted the cloud that had hung over her head for 40 years of having come from an "enemy" country. After 1949, the "Republic of China" became a strong ally of the United States in its global war with Communism; this was the main reason why Faina kept a low profile and avoided drawing attention to her origins. She rarely spoke Russian, except with her husband. She had thrown all her energy into being a good Chinese wife. Now her origins were no longer something to hide.

After the end of Soviet Union and the lifting of 70 years of Communist censorship, reports about her began to appear in the Russian press. Who was this Russian woman who had married CCK? How did she end up in Taiwan? No-one knew anything about her – how successful the censors had been! Newspapers interviewed her former friends and colleagues in Sverdlovsk. One still kept a photograph of her and said that she would like to receive her if she came home.

In the spring of 1992, out of the blue, Faina received a letter from one of her closest friends in Sverdlovsk, Maria Anikeyeva; she had attended Faina's marriage there in March 1935. This was the first letter the two had exchanged in 55 years. When they said goodbye that freezing day on the platform of the city's railway station, neither could have imagined that a World War and the Cold War that followed it would prevent a reunion. The courier was John Chang Hsiao-yan (蔣孝嚴), one of CCK's illegitimate sons with Chang Ya-juo; he had received the letter from a Russian professor. As a diplomat of the "Ministry of Foreign Affairs", he visited Russia in 1991 and 1992 after the collapse of the Soviet Union.

In the letter, dated February 23, 1992, Maria expressed her condolences for the loss of CCK. "How many children and grandchildren do you have? Is there anyone living with you? Do you have a pension to support your life?" She said that they had often seen Faina's elder sister Anna, who had shown them the letters and photographs she had sent. "Later, I do not know where she went and we lost touch. Everyone is very interested in your history. Many have visited our house. We showed the photographs you took on the Black Sea (during the honeymoon of CCK and Faina)."

Maria said that her husband Fyodor had died in 1988 and she had three sons, all married and working in Uralmash, as well as six grandchildren. From 1951 to 1955, her husband had led a team of engineers at a metallurgy plant in Anshan (鞍山) in northeast China and then spent a year in Beijing. "We very much wanted to see you but it was impossible. Please come to visit us and see Uralmash – it has greatly changed. Warmest blessings."

Eddie said that his mother was inhibited in her feelings toward the Soviet Union. "On the one side, she did not have any family there. On the other side, because of the political situation, she had to hide in her heart her memories of the Soviet Union. She never expressed them. Sometimes, when my parents were drinking, they talked in Russian, which we children did not understand. Sometimes they talked about the life they had in the Soviet Union. Otherwise, there was very little opportunity to see or hear her memories of the Soviet Union. Especially, in her old age, she became very quiet and could not easily express to others what was in her heart. The more she did not want to say, the more others did not dare to ask. They feared it would make her sad." (Wang Mei-yu, biography page 126).

On June 16, 1992, she received at Seven Seas a delegation from the city government of Minsk, the capital of Belarus. It was led by Mayor Alexander Gerasimenko and included Vice Mayor Pyotr Nikitenko. After the collapse of the Soviet Union, the new government of Belarus was free to conduct its own foreign relations. Minsk wanted to become a twin-city with Taipei. So it sent the delegation there for more than a week in June 1992. In an interview with the *Komsomolskaya Pravda* newspaper of Belarus in 2007, Nikitenko said that, while they were in Taipei, there was an opportunity to meet the widow of CCK. "We gasped when we learnt that she was Belarusian. We expressed a wish to meet her, to give her souvenirs. Our guides hesitated, saying that, after the death of her husband, she did not receive anyone. But then they called the visitors back in surprise. 'Madame Chiang is waiting,' they said. They added that she had declined a meeting with the Mayor of Moscow."

Faina asked to be accompanied by Nancy Hsu (徐乃錦), her eldest daughter-in-law, and James Soong Chu-yu (宋楚瑜),secretary-general of the Kuomintang party (國民黨秘書長); for a long time, he had served as English secretary to CCK. "The Foreign Ministry" also sent a department chief to attend. It was a rare opportunity for Faina to speak Russian. She was in good spirits and asked the visitors many questions about the new country of Belarus. "I have not returned for 55 years," she said.

"After the death of her husband, she had become an ordinary pensioner," said Nikitenko. "She lived in an independent two-storey house. Its décor was tasteful. Everything was stylish and discreet. We were greeted very warmly. The table was laid, with cookies, juice, tea, coffee. We planned to stay for about 20 minutes and talked for more than an hour. Despite her old age, she looked elegant. She was dressed in a simple, light-fitting dress. It was immediately apparent that she was the first lady.

"As natives of Belarus, we gave her a linen doll, Narochansky (rye) bread (a specialty of Minsk) and a national flag. She was very touched! In return, she presented us with national Chinese souvenirs – wicker reed mats with drawings and calligraphy. She said that she came from Orsha and asked about changes in the country and the formation of an independent state. We told her what our country is today, about the formation of an independent state and sovereignty. She told us of her grief. She had lost her two sons. We were warned before the meeting that it was undesirable to hold conversations on personal topics. We did not ask. She told us what she wanted to say."

The guests told her that they welcomed her to visit her motherland and that they would be delighted to receive such a distinguished compatriot. Nikitenko, an economist, is now a member of the National Academy of Sciences of Belarus. This was the only official contact she had with her homeland since she left the Soviet Union in 1937.

With the collapse of the Soviet Union, it became possible for her to go to Russia or Belarus for a visit. But she did not express such a wish. Her son Eddie said that he and his wife would take her on such a trip; but she did not ask for it. As she grew older, her health deteriorated and such a long journey became out of the question. "We offered to accompany her to Russia but she said that she did not want to go, because she had no relatives there. She suffered from asthma and had to wear a mask," said Elizabeth Chiang (interview with author in Taipei 19/12/2018).

Move to California

With the passing of her two elder sons, Faina found herself with no children living in Taiwan. Her daughter Amy had long settled in California; in 1988, Eddie had moved to Montreal and then San Francisco. He returned once a month to Taipei to see her. In 1991, seeing how traumatised she was by the loss of her other sons, he decided that it might be better for her to move to California, where he and his family and his sister could look after her. He found it hard to persuade her to leave the house to which she was so attached; she was also nervous of the fuss such a visit would cause in the media. Finally, she agreed to go to California in September 1992.

"I told Mother the purpose was to visit my sister and her grandchildren," said Eddie. "Actually, if she could adapt to the life there, I wanted her to settle with us there. (Wang Mei-yu biography, page 140). He kept the visit secret, to prevent her becoming the target of the news media.

In leaving Taiwan, Faina carried an ordinary passport and asked for no special diplomat treatment; she was entitled to it as the widow of a "president". The Leader Lee Teng-hui (李登輝) wanted to see her off, as a matter of courtesy; but she insisted he not go, because she did not wish to bother such a busy man. Nonetheless, Lee's wife and many senior officials went to the airport; it was a mark of respect to her late husband and to Faina herself. She was accompanied on the flight to California by son Eddie, his wife and a bodyguard. It was her second visit to the U.S.; her first was in 1969, for the wedding of son Alex, also in California. By contrast, when her mother-in-law SML moved to the U.S., she took 97 suitcases and carried a special passport. The contrast in style between the two Madame Chiangs could not have been sharper.

In San Francisco, she stayed in Eddie's home. Her daughter Amy came to visit her with her husband and son. It was the longest time Faina had spent with Amy since her marriage, which Faina and CCK had initially opposed. Eddie took Faina to the Golden Gate Bridge and other famous sites in the city. The weather was fine and the air was clear, which was good for Faina's asthma. She enjoyed everything she saw and her health improved.

Many Taiwan people have emigrated to San Francisco because of its fine location overlooking the Pacific Ocean, clean air, cool summers, rolling hills, high living standards and rich cultural life. These were some of the reasons for Faina to settle in San Francisco. She had two children and their families living there; they could provide her with every comfort and convenience and the emotional support of the family. The U.S. government would welcome her. But she decided not to stay because she wished to return to Taipei.

This is how her daughter-in-law, Elizabeth Chiang, explained it. "In 1992, Faina came to San Francisco, where I was living with our family, as was her daughter Amy. She spent a month there and was very happy. Then she said that she had to return to Taiwan because she had been there too long. 'Taiwan is my home. I must be close to my husband,' she said. Her doctor said that it would be better for her to live in a drier climate, such as in San Francisco." (Interview with author in Taipei, 19/12/2018).

This was a remarkable statement of love for her husband; she preferred to stay in the house where they had lived together than with her children in San Francisco. On October 20, she returned to Taiwan, accompanied by Eddie; she never left the island again.

Back to Seven Seas

She returned to Taipei and resumed her quiet life in Seven Seas. Her main companions were housekeeper Ah Bao, the nurse Miss Chen and members of her family, especially her daughters-in-law. Eddie recalled: "Each day she sat in Father's room, looking at all the items there that had not been moved, treasuring her memories, as if he was still there. This feeling could not be reproduced anywhere in San Francisco, however beautiful the conditions were." (Wang Mei-yu, biography page 145).

She rarely left the house, even to travel within Taiwan, and was sad that the grandchildren she loved did not live close to her; they had made new lives for themselves overseas. Faina read books and newspapers and watched television; she took regular walks in the garden and looked through family photograph albums.

In October 1994, she suffered a stroke and was rushed to VGH. The doctors said it was very dangerous; for a period, she could not speak. But her will was strong and, thanks to the excellent care of the doctors and nurses, she recovered and returned home. Nearly 80, she still had a strong will to live. But the angel of death had not finished stalking her family.

Passing of Eddie

In January 1996, doctors at the VGH diagnosed Eddie with cancer of the oesophagus (食道癌). After an operation, they told the family the disease was in the late stages and that they were not optimistic. They gave him three-to-six months to live. On the day of the operation, he ordered the staff at Seven Seas to remove all the newspapers; he knew that Faina read them each day and he did not want her to find out. "I really did not know how she would continue after I was gone," he said (Wang Mei-yu, biography page 149).

Like Faina and his own wife Elizabeth, Eddie was a devout Christian. He stopped drinking and smoking and underwent chemotherapy at VGH. When possible, he went to Seven Seas to have lunch with his mother. She regularly visited him in hospital. As his condition worsened and he became thinner and thinner, he knew he could not hide the illness from her any more. When he told her, he realised that she knew it already. In the summer of 1996, after the end of the chemotherapy, he went to the mainland to seek treatment from Chinese doctors. But it was too late; the cancer had already spread to other parts of the body. In the second half of the year, he divided his time between Taiwan and the U.S., where his children were studying and his grandmother (SML) lived; she would be 100 in the spring of 1997. In December 1996, he returned to Taiwan with his wife and three children.

It was December 22. His wife Elizabeth recalled: "That day we had lunch with our sons. When Eddie was taking a nap, the nurse came out to say that something was amiss with my husband. The doctor came and told us to prepare ourselves. I was there with my children at his side. Faina came to the hospital, held his hand. After he heard her speak, then he left us. He had to wait until he heard her voice. After his passing, Faina was very important to me. She gave me guidance." (Interview with author, Taipei, 19/12/2018)

Eddie was just 49. Faina had lost all three of her sons. It was the end of the "Chiang dynasty"; none of the grandchildren wanted a life in politics or in public life in Taiwan.

Tragedy or Retribution

Ask Taiwan people about this tragic story of a mother losing her three sons during her lifetime and they give different answers. Everyone expresses sympathy for Faina, whom they have seen on television or in the newspapers visiting the hospital or attending the funerals. They feel for her as a mother enduring this unspeakable loss. But many are less charitable toward the Chiang family. But many ordinary Taiwan people who spoke to the author in 2019 were less charitable toward the Chiang family.

"The death of Faina's three sons at a young age was 'bao ying' (報應),

retribution against CKS for all the Taiwanese he killed," said Lin Bei-lai (林北來), a high school teacher. "It was not against CCK, who was a better man than his father. The three sons of Faina died. Then it was Winston Chang Hsiao-tzu (章 孝 慈), Chancellor of Soochow University (東吳大學). When he was in the mainland, he fell ill and died at the age of 53. This was also part of the retribution."

Winston was one of the twins born to CCK and Chang Ya-juo in 1942. He died in VGH on February 24, 1996, after suffering a stroke.

"Many people in Taiwan hate the Chiang family," said Chang Mei-hwa (張美華), an office worker. "They killed so many. So it is that people have the idea of retribution against the family taken on the children, even if they themselves were not guilty. The history of the Chiang family is very dark. Faina herself was innocent. We know little about her. She spoke very little – perhaps she was not allowed to speak. She was a simple housewife, with no involvement in politics."

Huang Li-guo (黃 立 國), a Taipei taxi driver, mentioned the bad behaviour of Alan and Alex. "Faina's two eldest sons misbehaved. They drank, chased women and abused their bodies. So they became sick. They had privileges and no restraint. Why did their parents not control them? The Chiang family was like that. So, it was not a matter of destiny but a result of their own behaviour. It was very sad for Faina but not a divine revenge. CCK cared about people and worked for them. He was popular.

"Faina has a good reputation among the public. She was a wife, mother and housewife. She lived simply and was not corrupt. After her husband died, she stayed in Taiwan, which shows that she had friends here. As soon as CKS died, his wife went to live in the United States, where she had a fortune. Faina did not do that," Huang said.

Main sources for this chapter:

Memoirs of Chou Lian-hua (United Literature Publishing Company, Taipei, first edition 1994, second edition September 2016).

Interview with Madame Elizabeth Chiang (方智怡), Taipei, 19/12/2018.

Pyotr Nikitenko interview with the *Komsomolskaya Pravda* of Belarus in 2007.

The Generalissimo's Son, by Jay Taylor (Harvard University Press 2000).

Biography of Chiang Fang-liang, by Wang Mei-yu (China Times Publishing Company of Taiwan, 1997).

My Years at the Side of Chiang Kai-shek and his Son, by Weng Yuan (China Times Publishing Company of Taiwan, 2015).

Chiang Fang-liang and Chiang Ching-kuo, by Zhou Yu-kou (Rye Field Publishing Company, 1993).

CHAPTER 8

Passing and Legacy

After the passing of Eddie, Faina continued to live at Seven Seas. Her main companions were the staff and her three daughters-in-law. On January 13, 1997, the four of them went to Touliao to pay their respects (謁靈) at the grave of CCK on the ninth anniversary of his death. Faina lived a quiet life and went out little. She was sustained by her Christian faith and the memories of her 53 years of marriage.

Daughter-in-law Elizabeth Chiang (蔣方智怡)said: "In her later years, she was lonely of course. We knew that you must wait until God asks fo r you. Her brain remained very clear until the end. She was physically strong. She was a tough Chinese woman. She considered herself a Chinese. I respected her greatly. She did not give me rules or control me. My husband said that she never said 'No'. We very much looked up to her and CCK and CKS and SML.

"I also remember the words of Madame Soong Mei-ling in 1971 after the 'Republic of China' was expelled from the United Nations in a speech to a women's group. She also used words from the Bible – 2nd Corinthians, Chapter 4, verse 7-10: 'We have this treasure in jars of

Handwriting of Chiang Kai-shek, a gift for Faina's 50th birthday (Courtesy of "Academia Historica")

clay to show that this all-surpassing power is from God and not from us. We are hard pressed on every side, but not crushed; perplexed, but not in despair; persecuted, but not abandoned; struck down, but not destroyed. We always carry around in our body the death of Jesus, so that the life of Jesus may also be revealed in our body.'" (Interview with author, Taipei, 19/12/2018)

The Chiang family used these words to sustain them during the years when Taiwan was losing its international status to People's Republic of China. Faina also used them to sustain herself in the face of the tragedies in her family.

Passing

On October 6, 2004, she checked into the VGH, suffering from acute asthma and inflammation of the oesophagus. Her condition stabilised after treatment. But, just before midday on December 15, she experienced a sudden drop in blood pressure and ventricular fibrillation, according to the hospital's chief of surgery Lee Shou-tung (李壽東). "The medical team's efforts at resuscitation unfortunately failed," he said. She passed away at 12:40pm that day, of respiratory and cardiac failure. She was 88.

That afternoon "President" Chen Shui-bian (陳水扁) went to the hospital to express sympathy to the family members. "She displayed the virtues of a traditional Chinese woman – she was a good mother and a good wife who always put her family before everything else," Chen said. "Vice President" Annette Lu (呂秀蓮) said that Faina's "admirable, womanly character" would be long remembered.

Her funeral was held at the VGH on December 27, with the leaders of both major political parties in attendance. They included Chen Shui-bian and Annette Lu, as well as Ma Ying-jeou (馬英九) and Wang Jin-ping (王金平) of the Kuomintang. According to her wishes, she was cremated and her ashes placed next to the tomb of her husband in Touliao.

The government plans to turn their Seven Seas home into a cultural park, including the residence and a lake, with an area of 4.6 hectares. It will include a library with documents related to CCK. The first floor will show the reception rooms for guests and family gatherings and the second floor will show the bedrooms and office space. The government wants the park to give the public a deeper understanding of CCK's life.

The legacy of Chiang Ching-kuo

Chiang Ching-kuo has left an extraordinary legacy. In 2019, 31 years after his death, Taiwan has become one of the wealthiest places in Asia, with a boisterous democracy, a free media and a dynamic civil society. In 2017, it had a GDP of US$573 billion, ranking 22nd in the world, with a PPP per capita GDP of US$49,827, ranking 21st (Source: Taiwan government website). It is a major player in the world information and communication technology industry. In 2017, it was the world's 16th largest exporter and 18th largest importer of merchandise, according to the World Trade Organisation. At the end of March 2019, its foreign exchange reserves were a record US$464.08 billion, ranking fifth in the world, after mainland China, Japan, Switzerland and Saudi Arabia.

The end of martial law led to an explosion of civil society. Taiwan now has more than 2,500 civic associations, with 250,000 local groups or committees and a membership of more than 10 million people; they are involved in charity, sport, voluntary work, academic study, education, publishing and other fields.

It has four major Buddhist institutions. One of them, the Tzu Chi Foundation (慈濟基金會), is the largest non-government organisation in the Chinese world, with 10 million members worldwide, half of them in Taiwan and half overseas. Its income comes from public donations. At home, it runs schools, universities, hospitals, charity work and 4,500 recycling centres. Abroad, it has provided disaster relief to people in over 85 countries.

It delivers food, clothing, medicines, daily necessities and blankets made of recycled PET (polyethylene terephthalate) bottles to the victims; it runs free medical clinics and builds homes for those who have lost theirs. It operates the largest bone marrow bank in Asia. In Istanbul, it has established schools and a medical clinic for Syrians in exile, with the refugees themselves providing the staff, the students and the patients. It is a remarkable example of the power for good of ordinary people and what civil society can achieve.

Lee Teng-hui, the man CCK chose to succeed him, held Taiwan's first "presidential" election in 1996, which he won with 54 per cent of the vote. Such elections have been held every four years since. In 2000, Chen Shui-bian, candidate of the opposition Democratic Progressive Party, won with 39.3 per cent of the vote; he became the first DPP "president". Since then, the DPP and KMT have shared occupancy.

None of this would have happened without the reforms enacted by CCK, especially those in the last year of his life. While CCK would have disapproved of many DPP policies, he would be happy that the

Taiwan people have been able to choose their own government, at the national and local levels. On November 24, 2018, the island held local elections. On the same day, people were invited to vote on 10 referenda, including same-sex marriage. Such a referendum can be held if merely 1.5 per cent of the electorate – about 280,000 people – propose one. It is one of the lowest bars for public referenda in the world.

Diplomatically, CCK's greatest success was the Taiwan Relations Act (TRA) of April 1979 that was negotiated just after Washington recognised Beijing in 1978. Many people in Taiwan, and abroad, believed that, once the United States had reached a substantial level of economic and diplomatic engagement with People's Republic of China, it would abandon Taiwan, stop selling its arms and withdraw its political support. This has not happened.

On May 6, 2019, the American Institute in Taiwan (AIT) – its de facto embassy – moved to a new US$255-million hillside complex with 15,000 square metres in the Neihu (內湖) district of Taipei. It has nearly 500 staff, including active U.S. military personnel, working there. "It is a symbol of the strength and vibrancy of US-Taiwan partnership in the 21st century," said Marie Royce, U.S. Assistant Secretary for Educational and Cultural Affairs, during a dedication ceremony in June 2018. William Brent Christensen, AIT director, said:

"AIT hopes Taiwan will view this impressive structure as a long-term investment in our collective future. We look forward to moving in and getting to work on the next 40 years of U.S.-Taiwan cooperation."

In April 2019, former U.S. House of Representatives speaker Paul Ryan, a Republican, led a 26-member delegation to Taiwan, to attend an event that marked the 40th anniversary of the TRA. He served as House speaker – one of the most senior positions in the U.S. government – from October 2015 to January 2019. Among his delegation were four members of the House, and former and incumbent U.S. government officials. The U.S. has continued to sell Taiwan advanced weapons; in 2019, the two sides were negotiating the sale of 66 F-16V fighter jets for Taiwan.

In his speech at the TRA anniversary event, Ryan said that Taiwan's embrace of democracy "shows a better path for all Chinese people. Taiwan's leadership reflects an understanding of shared value that instills unique importance to the U.S.-Taiwan relationship – which, of course, stems from our shared values and our common embrace of democracy, free markets, rule of law, and human rights." If CCK were present in the room, he would have been delighted.

The most complex and challenging issue for Taiwan remains relations with mainland China. In March 1973, after he had been recalled to the Communist Party's Central Committee (中央委員會), Deng Xiaoping (鄧小平) called for direct negotiations with Taipei on reunification. CCK rejected this offer from his former fellow student in Moscow. He also rejected other approaches, including an offer of "one country, two systems" from Deng in 1981, and opportunities to meet him. CCK believed that, the Nationalists could create on Taiwan a democratic and economic model for the Chinese nation which everyone would aspire to.

Like other leaders in Beijing, Deng realised that the best hope of reaching a deal with Taiwan lay with CCK. He was committed to the ideal of a "united China" and had the personal authority to make an agreement with the mainland. Those who succeeded CCK would not have such an authority. Deng was correct in his analysis. All that has happened in Taiwan since CCK's death has greatly complicated the possibility of reunification. The democratic reforms mean that now the public would have to approve such a momentous decision. But their overwhelming preference is for the status quo.

In a speech in January 2019, China's President Xi Jinping (習近平) made a similar offer of "one country, two systems". It was strongly rejected by Taiwan's leader and the public. So, the leaders of Taiwan have been left to square the circle – maintain good relations with their most important economic partner who wants reunification and answer to a public that refuses it.

Many Taiwan people are ambiguous about the reforms which CCK set in motion. They welcome the democracy, freedom of speech and space given to civil society. But they believe that the island's politics have become too confrontational, driven by party factions and personal rivalries; these make it hard to reach consensus and implement the required policies.

Many argue that the island's economy has paid a heavy price for this freedom and democratisation. Over the last 10 years up to the end of 2019 wages in Taiwan grew the slowest of the four Asian Tigers – the others are Hong Kong, South Korea and Singapore. Thousands of young people cannot buy homes in the major cities. A million Taiwanese live and work in mainland China, mainly because of higher wages. Some are nostalgic for the era of CCK; they say the government then was able to make long-term economic plans and carry them out. Taiwan's geographic position next to mainland China and Beijing's claim to it greatly complicates its politics and diplomacy.

The judgement of most Taiwan people on CCK is positive. He concentrated the work of his government on building the economy and infrastructure of Taiwan. He was the first Taiwan leader to engage earnestly with and listen to the public, an example which later politicians have followed. He was the person who turned a one-party state under martial law into an open and free society.

Throughout CCK's life, Faina was his strongest supporter, especially in times of danger and uncertainty. She was at his side in the fear and poverty of Sverdlovsk, the terrifying raids of the Japanese bombers in Chongqing, the defeat in the civil war and making a new life in Taiwan. She was his constant companion during the last years, when he was battling severe pain from diabetes, weak legs and failing eyesight. CCK made a historic transformation of Taiwan – only he and Faina know how much of this he owed to her.

CCK and Faina in old age

Main source for this chapter:

Interview with Elizabeth Chiang (Taipei, 19/12/2018).

Thanks and Acknowledgements

We must thank many for this book, especially the people of Taiwan. I went there first in the early summer of 1981, to study Mandarin (國語) and work as an editor at United Daily News (聯合報), one of the island's big newspapers. Everyone was polite and welcoming. I spent the first night in the home of a friend in Hsinchu (新竹), a city 87 kilometres southwest of Taipei. I was awoken the next morning at 4:30am by the sound of a military song. Looking out of the window, I saw soldiers running around the sports field in their base and singing a patriotic song. It was the era of martial law; men from 19 to 36 had to serve in the army for two years. Chiang Ching-kuo (蔣經國) was "president". He was everywhere – on television and billboards, in the newspapers and in the conversations of people. The government controlled the media; there was news of him every day.

I went to live in a student hostel in Taipei. The other students were both Taiwanese and foreign. One was a German-speaking Swiss; sometimes he met Chiang Wei-kuo (蔣緯國), the adopted brother of CCK, who also spoke German. In 1936, CKS sent him to a military academy in Munich; after graduation, he joined the Wehrmacht and commanded a Panzer unit during the occupation of Austria in 1938. He was promoted to Lieutenant of a Panzer unit and was waiting to join the invasion of Poland. Then China's War Ministry ordered him to the U.S. for further training and he left Europe.

The Taiwan students were polite and friendly to their foreign guests; living in a one-party state, they were reserved in what they said, especially if the talk turned to politics. No-one spoke of Faina; many did not know CCK had a foreign wife. At the language school, the teachers were mainlanders (外省人), members of families who had come to Taiwan in 1949 with CKS. Many had parents well placed in the government or the Kuomintang. They often talked of the former lives of their families in Shanghai, Beijing and Nanjing (南京), the capital until 1949. They too did not speak of Faina.

As time went on and friendships developed, people became less guarded. The private life of the Chiang family was a taboo in the official media but many spoke about it in private. It was hard not to, because the family had dominated the life of Taiwan since 1949. People did not have the information that became available after the end of martial law – but there were rumours and stories, especially on the bad behaviour of the two sons. In general, people spoke favourably about CCK; they said that he had done much to build the Taiwan economy and was a more responsive leader than his father.

People made jokes about the thick Ningbo accent of CCK and his father; when he talked on television, many needed to read the subtitles on the bottom of the screen to understand what he was saying. One popular joke concerned CCK's choice of a successor. He was about to make a major speech to announce the person – but his secretary still did not know who it was. So he followed his boss into the men's room and asked him through the door. CCK replied: " 你等一會兒 " (Ni deng yihui, Wait a moment). But, because of the heavy Ningbo accent, the secretary thought he heard " 李登輝 " (Lee Teng-hui). The point of the joke was that Lee was an unexpected choice – and, many in the Kuomintang thought, would only be a temporary leader.

My friends told me that CCK had a foreign wife – but they knew nothing about her; she rarely appeared in public. All these conversations built a picture of CCK and his family, even though it was incomplete and not always accurate. A complex man, he lived an extraordinary life; he was someone well worth writing about, I thought.

So, my thanks first to those teachers, friends and ordinary members of the public in Taiwan during the two and a half years I spent there; they were so kind and shared their insights and feelings. Since then, I have been able to return many times. I have always found the same level of humour and friendliness; with the end of martial law and lifting of censorship, people became better informed and more willing to express opinions without looking over their shoulder.

For this book in particular, I want to thank Elizabeth Chiang (蔣方智怡) and Karen Chien (簡靜惠) for their precious time. I interviewed Alex Chiang (蔣孝武) in Tokyo in August 1990. The "Academia Historica" (「國史館」) in Taipei graciously provided more than 30 images of Faina which the reader can enjoy. During our visits for research and interviews, David Wang (王寶裕) and his family were most helpful, driving us to many places and telling us many things we did not know. We also thank Paul Kuobong Chang (張國葆), a senior official of the "Ministry of Foreign Affairs", for his support and encouragement.

As the reader will observe, we have quoted liberally from several excellent books about the life of Faina and her family. I express my gratitude to the authors and admiration for their meticulous work–Jay Taylor (丁大衛), Wang Mei-yu (王美玉), Weng Yuan (翁元), Zhou Yu-kou (周玉蔻), Archibold Chien (簡錦錐) and

Pastor Chou Lian-hua (周聯華牧師). The reader will find the titles and publishers in the notes to each chapter. We also thank Russian history scholar Dr Victor Zatsepine, of the Department of History of the University of Connecticut, for reading Chapter One and advising revisions and corrections, which we carried out. John Gittings, a China expert at The Guardian newspaper in Britain, did the same for three other chapters.

We thank the staff of Joint Publishing in Hong Kong for their unfailing support and great professionalism, especially Deputy Chief Editor Anne Lee (李安), Commissioning Editor Yuki Li (李毓琪) and editor Sisi Chang (張蘊之), as well as their colleagues who designed and laid out the book. We thank Donal Scully for his fine work in editing the English version. Norman Ching (程翰) did an excellent job of translation, as he has done for several of my books.

My beloved wife Louise accompanied me for all the travel and interviews in Taiwan and unstintingly guided the myopic foreigner stumbling through the Chinese forest.

Before writing this one, I did two books about foreigners who spent their adult life in China – Grandfather Frederick O'Neill, an Irish Presbyterian missionary in Manchuria from 1897 to 1942: and Sir Robert Hart, Director-General of the Qing dynasty Imperial Maritime

Customs from 1863 to 1911. The story of Faina inspired me in the same way. Like Grandfather and Sir Robert, she left her family and home country and threw herself into a new and unknown life. She had to learn the Chinese language, customs and manners and how to be a Chinese wife. She joined the most powerful family in the country and shared their triumphs, failures and tragedies.

I hope the reader finds her story as moving and dramatic as I do.

Mark O'Neill

China's Russian Princess: the Silent Wife of Chiang Ching-kuo

Author Mark O'Neill
Editor Donal Scully
Designer Alice Yim

Published by Joint Publishing (H.K.) Co., Ltd.
20/F., North Point Industrial Building, 499 King's Road, North Point, Hong Kong

Printed by Elegance Printing & Book Binding Co., Ltd.
Block A, 4/F., 6 Wing Yip Street, Kwun Tong, Kowloon, Hong Kong

Distributed by SUP Publishing Logistics (H.K.) Ltd
3/F., 36 Ting Lai Road, Tai Po, N.T., Hong Kong

First Published in March 2020
ISBN 978-962-04-4618-4

三聯書店
http://jointpublishing.com

JPBooks.Plus
http://jpbooks.plus